Publishing

Analyzing Receiver Operating Characteristic Curves *with* SAS®

Mithat Gönen

THE POWER TO KNOW®

The correct bibliographic citation for this manual is as follows: Gönen, Mithat. 2007. *Analyzing Receiver Operating Characteristic Curves with SAS®*. Cary, NC: SAS Institute Inc.

Analyzing Receiver Operating Characteristic Curves with SAS®

Contents

Preface

When I received my Ph.D. in statistics, I had not heard the term *receiver operating characteristic (ROC) curve*. Like most biostatisticians, my introduction to ROC curves came through my involvement in studies of medical diagnostics. Later, I realized that they are used, although not as much as they should be, in other fields, such as credit scoring and weather forecasting, which make intensive use of statistical prediction methods. As my exposure grew, I needed more options in SAS to analyze ROC curves. However, over time, I realized that SAS offers more on ROC curves than I initially thought. I also accumulated a good deal of statistical knowledge and SAS tricks on ROC curves that may be helpful for those who now find themselves where I found myself several years ago. I wrote this book to share these experiences.

There are two recent statistical texts that cover ROC curves: *Statistical Methods in Diagnostic Medicine* (Zhou, McClish, and Obuchowski, 2002) and *The Statistical Evaluation of Medical Tests for Classification and Prediction* (Pepe, 2003). There is a natural overlap of coverage between these two texts and this book, but there are also important differences. First and foremost, this book is for the practitioner and, as such, matters of implementation and practice take priority over theoretical concerns. I also made a conscious effort to cover other fields that benefit from ROC curves. Nevertheless, diagnostic medicine remains the most common area of ROC curve application and served in this book as the source of many examples.

It is not possible to write a book without excluding subjects that some readers would have found absolutely essential. I am happy to receive e-mail on these topics and willing to share whatever expertise and programs I may have.

It is also not possible to write a book without making mistakes. Despite my best efforts, you may find technical errors; please let me know about them. Also, refer to the companion Web site for this book at support.sas.com/gonen for updates to both the SAS code and the book.

ROC on.

Acknowledgments

This book benefited greatly from the careful review and suggestions of Eugenia Bastos, Zoran Bursac, David Olaleye, Paul Savarese, Barbara Schneider, Yu Shyr, David Schlotzhauer, and Doug Wielenga. Special thanks to Nicole Ishill at Memorial Sloan-Kettering for reviewing text and testing macros.

I am grateful to Judy Whatley, my editor, for her unfailing help and patience at every stage of the long process that resulted in this book. The following individuals at SAS Press were also instrumental in making this book possible: Mary Beth Steinbach, Kathy Restivo, Candy Farrell, Patrice Cherry, Jennifer Dilley, Shelly Goodin, and Liz Villani. Jeanne Ferneyhough of SAS Institute provided much-needed expert help with SAS/GRAPH.

Steven Larson, Heiko Schöder, Timothy Akhurst, Robert Downey, Richard Wong, John Carrino, and Andre Güettler generously gave permission to use their data sets as examples.

It would not have been possible to undertake such a project without the support and understanding of my wife, Elza, and my daughters, Deniz and Selin. They admirably endured my irregular schedules, attitudes, and mood swings.

x

Introduction

1.1 About Receiver Operating Characteristic Curves

This book describes how to analyze receiver operating characteristic (ROC) curves using SAS software. A *receiver operating characteristic curve* is a statistical tool to assess the accuracy of predictions. It is often abbreviated as ROC curve or ROC chart, the latter being used more often in data mining literature.

Making predictions has become an essential part of every business enterprise and scientific field of inquiry. A simple example that has irreversibly penetrated daily life is the weather forecast. Almost all news sources, including newspaper, radio, and television news, provide detailed weather forecasts. There is even a dedicated television channel for weather forecasts in the United States.

Of course, the influence of a weather forecast goes beyond a city dweller's decision to pack an umbrella. Inclement weather has negative effects on many vital activities such as transportation, agriculture, and construction. For this reason, collecting data that help forecast weather conditions and building statistical models to produce forecasts from these data have become major industries.

It is important for the consumers of these forecasts to know their accuracy. This helps them to incorporate these predictions into their future plans. It also helps them decide between competing providers. Similarly, it is important for forecast providers to assess the accuracy of their forecasts since accuracy is a direct indicator of the quality of their product. Assessing accuracy is also important when providers decide to invest in technologies to improve the forecasts. An improvement in the forecast is intrinsically linked to an improvement in accuracy.

Credit scoring is another example of making predictions: When a potential debtor asks for credit, creditors assess the likelihood of default to decide whether to loan the funds and at what interest rate. Accurately assessing a debtor's chance of default plays a crucial role for creditors to stay competitive. For this reason, the prediction models behind credit scoring systems remain proprietary. Nevertheless, their predictive power needs to be continuously assessed for creditors to remain profitable.

A final example concerns the field of medical diagnostics. The word *prediction* rarely appears in this literature, but a diagnosis is a prediction of what might be wrong with a patient exhibiting certain symptoms and complaints. Most diseases elicit a response that increases levels of a substance in the blood or urine. However, there might be other reasons for such elevated levels; hence, assessing blood or urine levels alone can create a misdiagnosis. It is critical to understand how accurate such diagnoses are because they influence subsequent evaluations and treatments. It is also common to have multiple diagnostic markers or tools available, and a fair assessment of them involves comparing their accuracies. These diagnostic options may differ in their cost and risk to the patient, in which case a decision analysis can be performed where the value of each tool is quantified based on the accuracy of the diagnoses made by the tool.

All these examples have a common theme. A prediction is made before the value of the entity that is predicted is known. We need a method to evaluate the accuracy of these predictions. As these examples make clear, it would be helpful if the method could compare the accuracy of several predictions.

Various conventions are used to name the predictions and the outcome. Table 1.1 summarizes the most commonly used names.

Table 1.1 Common Nomenclature for the Elements of an ROC Curve

Variable That Predicts	Variable to Be Predicted	Values of the Variable to Be Predicted
Predictor	Outcome	Case/Control
Marker	Status	Diseased/Non-Diseased
Score	Gold Standard	Positive/Negative
Forecast	Indicator	Present/Absent
		Event/Non-Event

ROC curves provide a comprehensive and visually attractive way to summarize the accuracy of predictions. They are widely applicable, regardless of the source of predictions. You can also compare the accuracy of different methods of generating predictions by comparing the ROC curves of the resulting predictions. Therefore, it may come as a surprise to realize that ROC curves are generally ignored during the education and training of statisticians. Most statisticians learn about ROC curves on the job, as needed, and struggle through some of the unusual features of this type of analysis.

To make matters worse for SAS users, very few direct methods are available for performing an ROC analysis in SAS. However, many existing procedures can be tailored with little effort to produce ROC curves. SAS Institute also provides a macro to perform some of the calculations. This book describes how to produce ROC curves with the available features in SAS and expands on further analyses using other SAS procedures.

1.2 Summary of Chapters

Methods for evaluation of accuracy depend on the nature of the predictor. Chapter 2, "Single Binary Predictor," and Chapter 3, "Single Continuous Predictor," introduce appropriate methods for binary and continuous variables. These two chapters discuss material that is used repeatedly in subsequent chapters, so you must have a good grasp of these concepts before reading further. If you are already familiar with these statistical concepts but are more interested in learning the capabilities of SAS with respect to ROC curves, skip the parts introducing and discussing these concepts. Most of the SAS code in this book is presented within the context of examples, so it will be sufficient for those readers to have a cursory reading of Chapters 2 and 3 to familiarize themselves with notation and then carefully follow the examples to master the SAS code.

Most of the computations are performed using PROC FREQ, PROC LOGISTIC, or PROC NLMIXED. There are also a few macros that are very useful in plotting the ROC curve or computing the standard errors of the areas under the ROC curves. Occasional calls to PROC TRANSREG (for Box-Cox transformation) or PROC MIXED along with the use of PROC SURVEYSELECT for creating bootstrap samples are used.

Note: There is no standard mathematical notation for most of what needs to be presented here. I tried to balance my personal preference with widely accepted practices; this is why a cursory reading is recommended, even for those who feel comfortable with the statistical concepts.

Chapter 4, "Comparison and Covariate Adjustment of ROC Curves," compares the ROC curves of several markers and adjusts them for covariates. The principle tool for this purpose is regression, which accommodates both categorical and continuous covariates. Regression methods can also be used to compare the accuracy of several markers by representing the markers with dummy variables in an ANOVA-like model. Although the mechanistic aspects of these regression models are similar to other regression models, the inclusion and interpretation of model coefficients are unique to the field of ROC curves.

Chapter 5, "Ordinal Predictors," repeats the material in Chapters 3 and 4 for an ordinal predictor. The ideas are very similar, but the statistical techniques are slightly different, such as the use of a latent variable probit regression model, which is also commonly called the *binormal model* in ROC literature. While it is possible to study Chapter 5 with only a superficial understanding of the earlier material, I recommend mastering the concepts in Chapters 2 and 3 first.

Chapter 6, "Lehmann Family of ROC Curves," and Chapter 7, "ROC Curves with Censored Data," present relatively new material that has not yet made its way into other books. The Lehmann family of ROC curves, the focus of Chapter 6, uses the proportional hazards model exclusively. Proportional hazards models are routinely used in survival analysis but rarely in other applications. Chapter 6 shows how it can be used to create ROC curves and extended to regression models and clustered data using the capabilities of PROC PHREG. Most statisticians identify PROC PHREG with censored data, but Chapter 6 deals with a binary outcome that is fully observed, just like the outcome variables in Chapters 2 through 5. The problem of creating ROC curves with censored data is tackled in Chapter 7. Two methods of computing a concordance probability are provided, along with a discussion of time-dependent ROC curves.

Chapter 8, "Using the ROC Curve to Evaluate Multivariable Prediction Models," discusses the use of ROC curves when multivariable prediction models are built and assessed on the same data set. Chapter 9 uses the same concepts in the context of data mining. Although the concepts remain the same, the primary SAS data mining engine, SAS Enterprise Miner, has a very different user interface and functionality than SAS/STAT software. Hence, most of Chapter 9 discusses how models are developed in SAS Enterprise Miner and how ROC curves can be produced using the built-in functionality for model assessment. Also shown are ways of exporting the data for the ROC curves so that you can create custom plots using SAS/GRAPH software.

Most of the SAS code presented consists of SAS/STAT and SAS/GRAPH software; Base SAS is used occasionally to prepare the data for analysis and plotting. SAS/STAT procedures FREQ, LOGISTIC, MIXED, and NLMIXED provide most of the required ammunition for the analyses.

You should have a basic understanding of linear models and regression analysis. This book assumes no prior experience with SAS/STAT procedures; however, if you aren't familiar with common SAS System concepts, such as BY processing or the CLASS statement, you may benefit from consulting a general-purpose SAS manual.

This book features a generic macro that can be used to plot ROC curves regardless of the nature or origin of the predictions. Those who find the options offered by this macro sufficient may not need any direct use of SAS/GRAPH software. However, graphical presentation involves a degree of personal style, and you might like to customize your curves and to use particular annotations. If so, you can use the intermediate data sets created by the macro and write your own SAS/GRAPH code to produce custom graphics.

It is likely that the SAS code presented in this book, especially the macros, will evolve. The code, which is available for download from the book's companion Web site at support.sas.com/gonen, will be updated routinely, so check the Web site frequently for the latest developments.

Single Binary Predictor

2.1 Introduction

One of the simplest scenarios for prediction is the case of a binary predictor. It is important not only because it contains the most critical building blocks of an ROC curve but also because it is often encountered in practice. This chapter uses an example from weather forecasting to illustrate the concepts. Problems dealing with similar data are abundant as well, ranging from diagnostic medicine to credit scoring.

2.2 Frost Forecast Example

Thornes and Stephenson (2001) reviewed the assessment of predictive accuracy from the perspective of weather forecast products. Their opening example is very simple and accessible to all data analysts regardless of their training in meteorology. The example discusses frost forecasts produced for the M62 motorway between Leeds and Hull in the United Kingdom during the winter of 1995. A *frost* occurs when the road temperature falls below 0 °C. First, the forecast for each night is produced as a binary indicator (frost or no frost). Then the actual surface temperature of the road is monitored throughout the night and the outcome is recorded as frost if the temperature dropped below 0 °C and as no frost if it did not drop below 0 °C. The guidelines provided by the Highways Agency mandate the reporting of results (both forecast and actual) in a consolidated manner (see the Frost and No Frost columns and rows in Table 2.1) only for the days for which the actual temperature was below 5 °C. The example refers to the winter of 1995, when the actual road surface temperature was below 5 °C on 77 nights. The results are given in Table 2.1. Such a tabular description is the standard way of reporting accuracy when both the prediction and the outcome are binary. It is visually appealing and simple to navigate, and it contains all the necessary information.

There were 29 nights when frost was forecast and a frost was observed, and there were 38 nights when no frost was forecast and no frost was observed. Those two cells (the shaded portion of Table 2.1) represent the two types of correct forecast. A general terminology for these two cells is *true positives* (*TP*) and *true negatives* (*TN*). The roots of this terminology can be found in medical diagnostic studies when a test is called positive if it shows disease and negative if it does not. By analogy, you can consider frost to mean positive and no frost to mean negative, in which case there are 29 true positives and 38 true negatives in Table 2.1.

Table 2.1 Forecast Accuracy for Road Surface Temperatures

Forecast	Observed		
	Frost	**No Frost**	**Total**
Frost	29	6	35
No Frost	4	38	42
Total	33	44	77

What about forecast errors? There were 6 nights when a frost was forecast and none was observed. There were 4 nights when no frost was forecast, but a frost was observed. One can easily extend the terminology to call these two cells *false positives* (*FP*) and *false negatives* (*FN*). Table 2.2 is a generic representation of Table 2.1 using the terminology introduced here.

Table 2.2 Reporting Accuracy for Binary Predictions

Forecast	Observed		
	Positive	**Negative**	**Total**
Positive	True Positive (TP)	False Positive (FP)	TP+FP
Negative	False Negative (FN)	True Negative (TN)	FN+TN
Total	TP+FN	FP+TN	TP+FP+FN+TN

2.3 Misclassification Rate

There are a variety of ways to summarize forecast accuracy. An obvious one is the *misclassification rate* (*MR*), which is the proportion of all misclassified nights, the sum of false negative and false positives, out of all nights:

$$MR = \frac{FN + FP}{TP + FN + FP + TN}$$

One minus the misclassification rate is sometimes called *percent correct* or simply *accuracy*. The MR for the data in Table 2.1 is 10/77=13%. = Misclass rate

While the misclassification rate is simple to compute and understand, it is sometimes too crude for understanding the mechanism behind misclassification. It is also prone to bias if the information is not assembled carefully. Suppose that instead of following the Highways Agency's guidelines, the forecast provider decided to include all nights in a calendar year. There are 77 nights reported in Table 2.1 and, by definition, all those nights the actual temperature dropped below 5 °C. Therefore, the remaining 288 nights were all above 5 °C (no frost), bringing column marginal totals to 33 nights with frost (unchanged) and 332 nights with no frost. It is possible that the MR for the 288 nights when the temperature was above 5 °C was much less than the MR for the nights when the temperature was below 5 °C. Suppose that the misclassification rate for these 288 nights was 5%, resulting in 15 misclassified nights (rounded up). Then there would be a total of 25 misclassified nights out of 365 and the MR would be 25/265=7%. Table 2.3 shows several possibilities.

Table 2.3 Percent Misclassified for the 288 Nights When the Temperature Was above 5 °C and the Corresponding MR for All 365 Nights

Percent misclassified for the 288 nights when the temperature was above 5 °C	0	1	2	4	6	8	10	12	13
Percent misclassified for all 365 nights	2.7	3.5	4.3	5.9	7.5	9.1	10.6	12.2	13.0

It is clear that the MR is sensitive to which nights are included in the sample because the performance of the forecast is not homogeneous for all the nights in a year. It is also clear that as you include more easy-to-forecast nights in the sample, the MR becomes smaller. You can safely assume that for warm days in spring and summer no nightly frost forecast for the M62 is necessary because most people can make the same prediction (no frost!) quite accurately without resorting to a scientifically obtained forecast. This explains why the Highways Agency restricts the accuracy reporting to nights when the actual temperature was 5 °C or less. It also highlights the fact that interpretation of the MR depends on the proportion of nights with frost in the sample (30% in this example). This proportion is sometimes called *prevalence*.

In unregulated areas such as credit scoring, where scoring algorithms remain mostly proprietary, there are no such rules, or even guidelines, on how accuracy should be evaluated. In addition, consumers of predictions are not always diligent or knowledgeable in interpreting the details or intricacies of accuracy. Therefore, you need measures that are more robust.

2.4 Sensitivity and Specificity

The most common way of reporting the accuracy of a binary prediction is by using the true (or false) positives and true (or false) negatives separately. This recognizes that a false negative prediction may have different consequences than a false positive one. It also makes these measures independent of prevalence. For this reason, these two measures are considered to gauge the inherent ability of the predictor. In this weather forecast example, a false positive is probably less costly because its primary consequence may be more cautious and better-prepared drivers. On the other hand, a false negative may end up in insufficient preparation and accidents. This suggests reporting false positive and false negative rates separately.

$$FPR = \frac{FP}{TN + FP}, \quad FNR = \frac{FN}{TP + FN}$$

It is more common to work with true positive and true negative rates, defined as

$$TPR = \frac{TP}{TP + FN}, \quad TNR = \frac{TN}{TN + FP}$$

The true positive rate (TPR) is sometimes called *sensitivity* and the true negative rate (TNR) is sometimes called *specificity*. While these are generic terms that routinely appear in statistics literature, each field has come up with its own terminology. Weather forecasters, for example, use the miss rate for FNR and the false alarm rate for FPR.

The FNR and FPR for the weather forecast data are 4/33=12% and 6/44 = 14%, respectively. The sensitivity and specificity are 29/33=88% and 38/44=86%, respectively. In this instance, the FPR and FNR are both very close to each other. When this is the case, they will also be very close to the MR. In fact, the MR is a weighted average of the FPR and the FNR: *MR = w*FPR + (1-w)*FNR*, where the weight (*w*) is the proportion of nights with an observed frost. This is sometimes called the *prevalence* of frost.

Note that the denominators for TPR and TNR are the total observed positives and negatives. It is possible to define similar quantities using the forecast positives and negatives as the denominator. In this case, the ratios corresponding to sensitivity and specificity are called the *positive predictive value* (*PPV*) and *negative predictive value* (*NPV*):

$$PPV = \frac{TP}{TP + FP}, \quad NPV = \frac{TN}{TN + FN}$$

SAS users familiar with PROC LOGISTIC recognize that PPV and NPV are actually called TPR and TNR. This may be a source of confusion. This book doesn't use the established PROC LOGISTIC terminology because it defines TPR and TNR consistent with the way they are used in medical diagnostics literature. Nevertheless, to minimize confusion, sensitivity and specificity are used more often than TPR and TNR.

2.5 Computations Using PROC FREQ

It is relatively easy to compute all of these measures in SAS. The following DATA step prepares the data for subsequent use by PROC FREQ:

```
data m62;
    input forecast $ observed $ weight;
    datalines;
Frost Frost 29
Frost No_Frost 6
No_Frost Frost 4
No_Frost No_Frost 38
;
run;
```

The following execution of PROC FREQ provides the necessary calculations:

```
proc freq data=m62;
    table forecast*observed;
    weight weight;
run;
```

Output 2.1 contains many useful pieces of information. As a quick refresher on PROC FREQ output, the key to the four numbers in each cell is found in the upper left portion of the table. The first number is the frequency (i.e., TP, FP, FN, and TN). The second number (Percent) uses the table sum (the sum of column sums or the sum of row sums, 77 in this example) as the denominator. The third number (Row Pct) is the row percentage (i.e., the proportion that uses row sums as the denominator). Finally, the fourth number (Col Pct) is the column percentage using column sums as the denominator.

Output 2.1

```
forecast      observed
Frequency|
Percent  |
Row Pct  |
Col Pct  | Frost  |No_Frost|  Total

Frost          29         6        35
            37.66      7.79     45.45
            82.86     17.14
            87.88     13.64

No_Frost        4        38        42
             5.19     49.35     54.55
             9.52     90.48
            12.12     86.36

Total          33        44        77
            42.86     57.14    100.00
```

Each of these numbers has a role in computing the measures discussed so far. For example, the MR is the sum of the table percentages (the second set of numbers) in the off-diagonal elements of the table: 7.79+5.19=12.98%. The PPV and NPV can be found among row percentages since row sums pertain to the predictions. Here, the PPV is 82.86% and the NPV is 90.48%. Finally, the sensitivity and specificity are available from the column percentages (87.88% and 86.36% in this example), implying a sensitivity of 87.88% and a specificity of 86.36%.

It is very important to understand the correct interpretation of sensitivity, specificity, PPV, and NPV. Let's start with predictive values first. Their denominators are the number of positive and negative forecasts. In the weather forecast example, the PPV can be interpreted as the probability of an actual frost when a frost is forecast, and the NPV is the probability of observing no frost when no frost is forecast. Hence, if a frost is forecast for the M62 motorway, the probability that there will actually be a frost is estimated to be 82.86%. Similarly, if the forecast does not call for a frost, then the probability that there will be no frost is estimated to be 90.48%.

In contrast, the denominators for sensitivity and specificity are observed positives and negatives. Therefore, sensitivity is the probability that a night with a frost will be correctly identified by the forecast out of all nights with a frost during the winter; similarly, specificity is the probability that a night without a frost will be correctly identified by the forecast out of all nights with no frost (and less than 5 °C) during the winter. In our example, the probability that a frost will be correctly forecast is estimated to be 87.88% and the probability that no frost will be correctly forecast is estimated to be 86.36%.

It is easy to imagine the utility of these probabilities as occasional consumers of a weather forecast. If you drive from Leeds to Hull on the M62 only a few nights during a winter, all you care about is whether the forecast will be accurate on those few nights. On the other hand, the agency responsible for keeping the motorway free of frost might be more interested in sensitivity and specificity when deciding whether to pay for these forecasts since that speaks to the entire "cohort" of nights in the upcoming winter, and the decision of the agency should be based on the performance of the forecast in the long run.

It is easy to recognize that sensitivity, specificity, and positive and negative predictive values are all binomial proportions if the corresponding denominators are considered fixed. The technical term is *conditioning* on the denominators. This gives easy rise to the use of binomial functionality within PROC FREQ to compute interval estimates.

The following code uses two separate calls to PROC FREQ to obtain estimates of sensitivity and specificity along with confidence intervals and a test of hypothesis about a specific null value. Note the use of the WHERE statement to choose the appropriate denominator for each calculation:

```
title "Sensitivity";
proc freq data=m62(where=(observed='Frost'));
    table forecast / binomial(p=0.8);
    weight weight;
    exact binomial;
run;
```

The values specified in parentheses following the BINOMIAL keyword are the null values for the test of hypothesis. A null value of 80% is specified. This value is chosen only to demonstrate the hypothesis testing feature of the BINOMIAL option. It does not correspond to an actual value of interest in the example.

Output 2.2 shows the sensitivity, which is estimated to be 87.88% or 29 out of 33. Each of these 33 nights of frost can be thought of as independent Bernoulli variates: 29 of them were positive (1) and 4 were negative (0). Their sum (29) is the binomial variate with a sample size (denominator) of 33, with sensitivity as the binomial proportion of the variate.

Output 2.2

```
Sensitivity

The FREQ Procedure

                                        Cumulative    Cumulative
forecast      Frequency      Percent     Frequency      Percent
─────────────────────────────────────────────────────────────────
Frost            29          87.88          29          87.88
No_Frost          4          12.12          33         100.00


           Binomial Proportion
            for forecast = Frost
       ──────────────────────────────
Proportion                  0.8788
ASE                         0.0568
95% Lower Conf Limit        0.7674
95% Upper Conf Limit        0.9901

Exact Conf Limits
95% Lower Conf Limit        0.7180
95% Upper Conf Limit        0.9660

   Test of H0: Proportion = 0.8

ASE under H0                0.0696
Z                           1.1315
One-sided Pr >  Z           0.1289
Two-sided Pr > |Z|          0.2578

Exact Test
One-sided Pr >=  P             0.1821
Two-sided = 2 * One-sided      0.3643

Sample Size = 33
```

The first set of confidence limits are based on the well-known normal approximation to the binomial. The key quantity for these confidence intervals is the *asymptotic standard error (ASE)*, which is given by

↪ binomial dist.

$$ASE = \sqrt{\frac{p(1-p)}{n}}$$

where n is the denominator for the binomial proportion (in this example, sensitivity) and p is the estimate of the proportion from the data. The ASE for sensitivity is 5.68%.

The confidence limits appearing under the ASE are based on asymptotic theory. If n is large, then the 95% confidence interval can be calculated using

$$p \pm 1.96 \times ASE$$

The output reports that the asymptotic 95% confidence interval for sensitivity is (76.74%, 99.01%).

Exact confidence limits, in contrast, are based on the binomial distribution, and they have better coverage in small samples and/or rare events. Because they are calculated by default when the BINOMIAL option is specified, there is no need to use the asymptotic confidence limits. The exact 95% confidence interval for sensitivity is (71.80%, 96.60%). The difference between the asymptotic and exact intervals highlights the typical effects of moderate sample size.

Finally, the last part of the output provides a test of whether the sensitivity is equal to 80% or not. The z-statistic reported in the PROC FREQ output is computed by $z = p/ASE(null)$, where $ASE(null)$ means the ASE under the null hypothesis. This is a typical way of computing test statistics, sometimes referred to as a *Wald test*. The $ASE(null)$ for our hypothesis is 6.96% and the corresponding z-value is 1.13. In large samples, z has a normal distribution under the null hypothesis so a p-value can be obtained by referring to a standard normal table. This results in a two-sided p-value of 0.26, suggesting that the sensitivity of frost forecast is no different than 80%.

There is an important difference between confidence intervals and hypothesis tests in general regarding the computation of the asymptotic standard error. The ASE of 5.68% uses the observed p of 87.88% while the ASE under H_0 (6.96% here) uses the p of 80% from the null hypothesis.

Output 2.3 pertains to specificity.

Output 2.3

```
Specificity

The FREQ Procedure

                                  Cumulative   Cumulative
forecast     Frequency   Percent   Frequency     Percent

No_Frost          38      86.36          38       86.36
Frost              6      13.64          44      100.00


         Binomial Proportion for
            forecast = No_Frost

Proportion (P)                0.8636
ASE                           0.0517
95% Lower Conf Limit          0.7622
95% Upper Conf Limit          0.9650

Exact Conf Limits
95% Lower Conf Limit          0.7265
95% Upper Conf Limit          0.9483

   Test of H0: Proportion = 0.75

ASE under H0                  0.0653
Z                             1.7408
One-sided Pr >  Z             0.0409
Two-sided Pr > |Z|            0.0817

Exact Test
One-sided Pr >=  P            0.0523
Two-sided = 2 * One-sided     0.1046

Sample Size = 44
```

The following code creates Output 2.3:

```
title "Specificity";
proc freq data=m62(where=(observed='No_Frost')) order=freq;
   table forecast / binomial(p=0.75);
   weight weight;
   exact binomial;
run;
```

The interpretation of the output for specificity is similar to that of sensitivity. Specificity is estimated to be 86.36% with an ASE of 5.17%. The 95% confidence interval based on the ASE is (76.22%, 96.50%) and the 95% exact confidence interval is (72.65%, 94.83%). A test of whether the specificity exceeds the pre-specified target of 75% yields an asymptotic *p*-value of 0.0817 and an exact *p*-value of 0.1046. You would retain the null hypothesis in this case based on this analysis.

The ORDER= option in PROC FREQ may be helpful in situations where SAS, by default, is choosing a binomial event different from the one you want. This leads to the computation of binomial proportions in a way that is the complement of the desired category. You can always subtract the values in the output from 1 and adjust the hypothesized value in a similar fashion to obtain the correct analysis. But it is also possible to obtain exactly the desired output by using the ORDER option of the PROC FREQ statement. To better understand the correct usage of the ORDER statement, it is important to understand that, by default (that is, in the absence of an ORDER option) PROC FREQ uses as event the first value in the alphabetically ordered list of unformatted values of the TABLE variable. This leads to frost being considered an event and the reported analysis is for FPR, not for specificity. ORDER=FREQ tells PROC FREQ to use the category that is most common as the event, leading to the desired analysis. This works when the specificity is greater than 50%, which should cover most cases. When the specificity is less than 50%, then ORDER=FREQ also uses the incorrect category for events. A fail-proof system requires sorting the data first by the predictor variable, in a descending fashion, and then using PROC FREQ without the ORDER= specification.

Single Continuous Predictor

3.1 Dichotomizing a Continuous Predictor

Now let's consider the problem that naturally leads to the use of the ROC curve. Suppose you are trying to predict a binary outcome, like the weather forecast, but instead of a binary predictor you have a continuous predictor. Most weather forecasts are produced by statistical models, which generate a probability level for the outcome. For example, most weather forecasts mention rain in their summary but if you look at the details they also report a chance of rain. How would you assess the predictive accuracy of these probabilities (or, perhaps more appropriately, the model that produced these predicted probabilities)?

Since you know how to analyze predictive accuracy for a binary predictor (using metrics such as sensitivity and specificity), you might consider transforming the predicted probability into a dichotomy by using a threshold. The results, however, would clearly depend on the choice of threshold. How about using several thresholds and reporting the results for each one? The ROC curve offers one way of doing this. The ROC curve offers one way of doing this by focusing only on sensitivity and (one minus) specificity.

Let's use an example from the field of medical diagnostics. One way to diagnose cancer is through the use of a special scan called *positron emission tomography* (*PET*). PET produces a measure called the *standardized uptake value* (*SUV*), which is a positive number that indicates the likelihood that the part of the body under consideration has cancer. After the SUV is measured, the patient undergoes a biopsy where a small piece of tissue from the suspected area is removed and examined under the microscope to evaluate whether it is cancerous or not. This process, called *pathological verification*, is considered the gold standard. Using the terminology introduced at Chapter 1, we will call the SUV the marker, and pathological verification the gold standard.

The data are reported by Wong et al. (2002). There are 181 patients, 67 of whom are found to have cancer by the gold standard. Because we are dealing with a continuous predictor, it is no longer possible to report the data in a tabular form. Instead you can use the following side-by-side histograms from PROC UNIVARIATE that can be obtained with the simultaneous use of CLASS and HISTOGRAM statements. For example, the following code was used to generate Figure 3.1:

```
proc univariate data=HeadNeck noprint;
  class gold_standard / keylevel='-';
  var suv;
  histogram suv / turnvlabels cfill=ligr cframe=white
    cframeside=white endpoints=0 to 20 by 2 cbarline=black
    font=swissb height=4;
run;
```

Figure 3.1 Side-By-Side Histograms for SUV

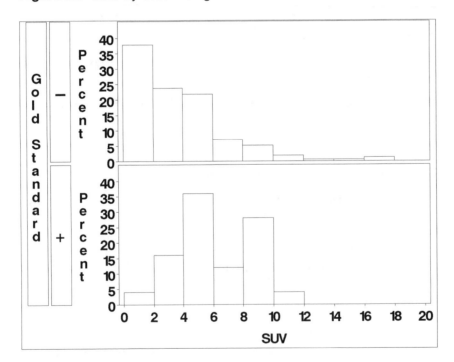

When the HISTOGRAM statement is invoked in PROC UNIVARIATE with a CLASS statement, a separate histogram for each value of the class variable is produced. The KEYLEVEL= option ensures that the chosen level always appears at the top (in this case, the code displays the histogram for the gold standard negative patients first). The options used in the HISTOGRAM statement visually enhance the figure. Their use is covered in detail in the *Base SAS Procedures Guide: Statistical Procedures*, which is available at support.sas.com/documentation/onlinedoc/sas9doc.html.

The upper panel of the figure is the histogram of SUV for the gold standard negative patients (those without cancer) and the lower panel is for gold standard positive patients (those with cancer), as denoted by the vertical axis. Patients with cancer tend to have higher SUV values; only a few have very low SUVs. A very high SUV (say, 10 or more) almost certainly implies cancer. There is, however, some overlap between the distributions for SUVs in the middle range (roughly 4–10). So extreme values of the SUV strongly imply cancer or lack of cancer, but there is a gray area in the middle. This is in fact a very common picture for many continuous predictors. How accurate, then, is SUV for diagnosing cancer?

One crude way of approaching the problem is to compute the sensitivity and specificity for various thresholds. Figure 3.2 uses a threshold of 7. To the left (SUV≤7) are considered negative for cancer and to the right (SUV>7) are considered positive for cancer. HREF=7 option is used in the HISTOGRAM statement for drawing the reference line and FRONTREF options was used to ensure that the reference line is in front of the histogram bars.

A greater proportion of gold standard negative patients have SUV≤7 and are classified as without cancer. Similarly, a greater proportion of gold standard positive patients have SUV>7 and are classified as having cancer.

Figure 3.2 Dichotomizing at SUV=7

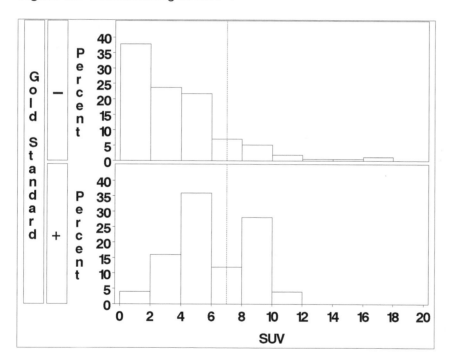

Dichotomization of the SUV can be accomplished using the following DATA step command:

```
suv7=(suv>7);
```

This can produce undesirable behavior if the SUV variable contains missing values because a missing value will satisfy suv≤7, which will be recoded as 0 in this case. IF-THEN statements should be employed if the variable to be recoded contains missing values.

Using the variable SUV7, you can easily obtain the kind of 2x2 table considered in the previous chapter for the weather forecast example. For this data set, it would look like the cells in Table 3.1:

Table 3.1 Accuracy of SUV>7 in Cancer Diagnosis

SUV>7 Classification	Gold Standard Diagnosis		
	Cancer	No Cancer	Total
Cancer	25	3	28
No Cancer	42	111	153
Total	67	114	181

The rows are classifications by the SUV>7 rule and the columns are classifications by the gold standard. Using the techniques covered in Chapter 2, you can use PROC FREQ to estimate the sensitivity and specificity as 25/67=37% and 111/114=97%. Therefore, using 7 as the SUV threshold to classify patients as having cancer or not is highly specific but of questionable sensitivity. This is not necessarily a verdict on the ability of the entire spectrum of standardized uptake values to diagnose cancer. In fact, Figure 3.2 suggests that moving the threshold to the left (i.e., using a lower threshold) will increase sensitivity at the cost of reducing specificity. Because the sensitivity using 7 as the threshold is so poor and the specificity so high, this might be a worthwhile strategy. This might help you find a better threshold, but it still falls short of determining whether the SUV is a good predictor of cancer. The next section explains how this can be done using ROC curves.

3.2 The ROC Curve

The previous analyses can be repeated for various thresholds, each of which may produce different values of sensitivity and specificity. One way to report the results of such an analysis would be in tabular form. Table 3.2 is an example, reporting on a few selected thresholds.

Table 3.2 Accuracy of SUV in Diagnosing Cancer for Various Thresholds

Threshold	1	3	5	7
Sensitivity	97%	93%	61%	37%
Specificity	48%	65%	88%	97%

Table 3.2 makes clear the wide range of sensitivity and specificity that can be obtained by varying the threshold. It also identifies the inverse relationship between the two measures: As one increases, the other decreases and vice versa. This can be seen from the histogram. If you move the dashed line to the left (to 5, for example, instead of 7), more patients will be classified as positive: Some of these will be gold standard positive, hence true positives, which will increase the sensitivity. Others, however, will be gold standard negative, and hence false positives, which will decrease specificity. So the relationship observed in Table 3.2 is universal: It is not possible to vary the threshold so that both specificity and sensitivity increase.

A tabular form is limited in the number of thresholds it can accommodate. You can plot sensitivity versus specificity, making it possible to accommodate all possible thresholds. Table 3.2 shows that since the two are inversely related, the plot of sensitivity against specificity will show a decreasing trend. A visually more appealing display can be obtained by plotting sensitivity against one minus specificity. This is called the *receiver operating characteristic* (*ROC*) curve. Figure 3.3 represents the ROC curve corresponding to the PET data set.

Figure 3.3 The ROC Curve for the PET Data Set

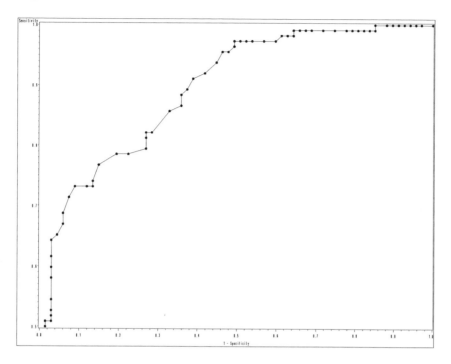

The dots are referred to as *observed operating points* because you can generate a binary marker that performs (or *operates*) at the sensitivity and specificity level of any of the dots. In this graph, the operating points are connected, leading to the so-called *empirical* ROC curve. The implication is that any point on the ROC curve is a feasible operating point, although you might have to interpolate between the observed marker values to find the correct threshold.

The ROC curve in Figure 3.3 is produced using the following code:

```
proc logistic data=HeadNeck noprint;
  model gold_standard=suv / outroc=ROCData;
run;

symbol1 v=dot i=join;
proc gplot data=ROCData;
  plot _sensit_*_1mspec_;

run;
quit;
```

The data set produced by the OUTROC= option in PROC LOGISTIC automatically computes sensitivity and one minus specificity for each possible threshold and names these variables _SENSIT_ and _1MSPEC_, which can then be plotted using standard SAS/GRAPH procedures.

Each point on the ROC curve corresponds to a threshold, although the value of the thresholds is not evident from the graph. This is considered a strength of ROC curves because it frees the evaluation of the strength of the marker from the scale of measurement. The ROC curve for the SUV measured on a different scale would be identical to the one produced here. Nevertheless, it may be helpful to indicate the threshold values at a few selected points to improve the understanding of the ROC curve. To highlight the sensitivity and specificity afforded by these thresholds, horizontal and vertical reference lines can also be included at selected points.

Inclusion of possible thresholds is not the only visual enhancement you can make. The most common enhancement is the inclusion of a 45-degree line. This line represents the ROC curve for a noninformative test and a visual lower bound for the ROC curve of interest to exceed. Finally, it is customary to make the horizontal and vertical axes the same length, resulting in a square plot, as opposed to the default rectangular plot of SAS/GRAPH. These features are available in the %PLOTROC macro. For more information, see this book's companion Web site at support.sas.com/gonen.

Figure 3.4 shows the ROC curve for the SUV with these enhancements. It was produced by %PLOTROC macro with the following call:

```
%plotroc(data=HeadNeck,marker=suv,gold=gold_standard,anno=4,tlist=3 4 5);
```

The %ROCPLOT macro, available from SAS, can be useful for this purpose as well.

Figure 3.4 The ROC Curve for the PET Data Set with Enhancements

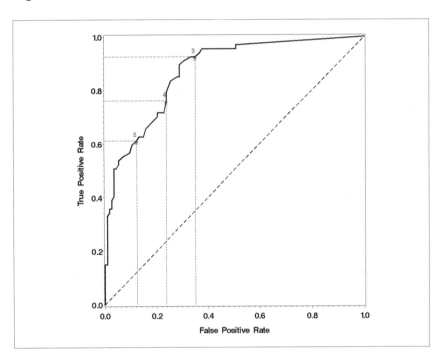

3.3 Empirical ROC Curve and the Conditional Distributions of the Marker

A convenient mathematical representation of the empirical ROC curve yields further insight into many of its properties. As Figure 3.1 shows, the distribution of the marker should be examined based on the gold standard value. Let $f(t|D=0)$ and $f(t|D=1)$ be the conditional density of the marker for gold standard negative and positive patients. The upper histogram is an approximation for $f(t|D=0)$ and the lower one for $f(t|D=1)$. Now define F and G as the survival functions (one minus the cumulative distribution) corresponding to $f(t|D=0)$ and $f(t|D=1)$, that is

$$F(t) = \int\limits_{t}^{\infty} f(s \mid D = 0)\, ds$$

$$G(t) = \int\limits_{t}^{\infty} f(s \mid D = 1)\, ds$$

Because all patients in *G(t)* are, by definition, gold standard positive, *G(t)* describes the proportion of positive patients whose SUV exceeds *t* out of all positive patients. This is nothing but the sensitivity when *t* is used as the threshold. Similarly, $1-F(t)$ would be the specificity; hence, *F(t)* represents one minus the specificity. Therefore, an ROC curve is a plot of $F(t)$ vs $G(t)$ for each *t*.

Although this is mathematically correct, it does not completely describe the ROC curve because of its dependence on *t*. However, we know that $F(t)$ is the x-coordinate of the ROC curve. Writing $x=F(t)$ and solving for *t*, you get $t=F^{-1}(x)$. The sensitivity corresponding to *t* is $G(t)$. So the sensitivity corresponding to *x* is given by

$$y=G(F^{-1}(x))$$

There are several important features of this representation. First of all, it is generic. No assumptions are made about the marker or the gold standard. It also makes explicit the dependence of the ROC curve on the entire distribution function. This representation also explains why the ROC curves focus on sensitivity and specificity instead of NPV and PPV. The denominators of the positive and negative predictive values change with the threshold and do not lend themselves to notation and analyses by the use of cumulative distribution functions of the predictor variable.

This relationship can also be used to highlight a fundamental property of the empirical ROC curve that it is invariant under monotone transformations of the marker. To understand this concept, imagine using *h(t)* instead of *t* as the marker, where *h(.)* is a monotone function (a *monotone function* preserves the ranks of the data; almost all of the transformations used in statistics, such as logarithmic or power, are monotone). *F* and *G* would have the same shape for *h(t)* as they do for *t*; the only difference would be in their horizontal axes. This would only change the thresholds, not the resulting ROC curve.

3.4 Area under the ROC Curve

The ROC curve is a summary of the information about the accuracy of a continuous predictor. Nevertheless, sometimes you might want to summarize the ROC curve itself. The most commonly used summary statistic for an ROC curve is the area under the ROC curve (AUC).

In an empirical ROC curve, you can estimate the AUC by the so-called *trapezoidal rule*:

1. Trapezoids are formed using the observed points as corners.
2. Areas of these trapezoids are calculated with the coordinates of the corner points.
3. These areas are added up.

This may be quite an effort for a curve like the one in Figure 3.3 with many possible thresholds. Fortunately, the AUC is connected to some well-known statistical measures, such as concordance, Somers' D, and the Mann-Whitney test statistic. This connection is associated with the invariance property discussed in Section 3.3. Invariance to monotone transformations implies that the ROC curve is a rank-based measure. The only information used from the observed marker values is their relative rank.

We will exploit these relationships not only to facilitate computation but also to gain further insight into the meaning of the area under the curve and to improve interpretation. In particular, we will see how you can estimate the concordance using PROC LOGISTIC and Somers' D using PROC FREQ. The relationship between the AUC and the Mann-Whitney statistic is as useful as the others in terms of conceptualization but not as useful from a SAS perspective. It is not explored here.

3.4.1 Concordance and Computing the AUC Using PROC LOGISTIC

Concordance probability measures how often predictions and outcomes are concordant. Continuing with the PET example, if, in a randomly selected pair of patients, the one with the higher SUV has cancer and the one with the lower SUV has no cancer, then this pair is said to be a *concordant pair*. A pair where the patient with the higher SUV has no cancer but the one with the lower SUV has cancer is said to be a *discordant pair*. Some pairs have the same SUV and they are called tied pairs. Finally, some pairs are noninformative; for example, both patients may have cancer or both may have no cancer. It is not possible to classify these pairs as concordant or discordant.

The probability of concordance is defined as the number of concordant pairs plus one-half the number of tied pairs divided by the number of all informative pairs (i.e., excluding non-informative pairs). In essence, each tied pair is counted as one-half discordant and one-half concordant. This is a fundamental concept in many rank procedures, sometimes referred to as *randomly breaking the ties*.

An equivalent way to express concordance is $P(SUV_+ > SUV_-)$, where SUV_- indicates the SUV of a gold standard negative patient and SUV_+ indicates the SUV of a gold standard positive patient. Figure 3.1 is helpful in visualizing this process. If one SUV is selected at random from the upper histogram and one from the lower histogram, what are the chances that the one chosen from the lower is higher than the one chosen from the upper?

The most important consequence for our purposes is that the area under the ROC curve is equal to the concordance probability that is reported by various SAS procedures. For example, PROC LOGISTIC reports concordance in its standard output under the heading "Association of Predicted Probabilities and Observed Responses." Using the same call to PROC LOGISTIC that generates the data set for plotting the ROC curves, you can obtain the AUC from the output marked with "c". In the PET example, this turns out to be 0.871 (or 87.1%), as seen from Output 3.1. The output also contains the elements of the computation (namely, the number of concordant, discordant, and tied pairs).

Output 3.1

```
Association of Predicted Probabilities and Observed Responses

Percent Concordant    86.1    Somers' D    0.743
Percent Discordant    11.8    Gamma        0.758
Percent Tied           2.1    Tau-a        0.348
Pairs                 7638    c            0.871
```

The AUC is estimated from the data, so there has to be a standard error associated with this estimation. The major drawback to using PROC LOGISTIC to estimate the AUC is that the associated standard error is not available. Fortunately, PROC FREQ provides an indirect alternative for this purpose.

3.4.2 Somers' D and AUC Using PROC FREQ

PROC LOGISTIC output includes Somers' D as well. D and AUC are related to one another through the equation $D=2*(AUC-0.5)$. Somers' D is simply a rescaled version of the AUC (or concordance) that takes values between -1 and 1, like a usual correlation coefficient, instead of 0 and 1. PROC LOGISTIC does not report the standard error for Somers' D. However, PROC FREQ reports both Somers' D and its standard error, as in the following example:

```
proc freq data=HeadNeck;
  table suv*gold_standard / noprint measures;
run;
```

The NOPRINT option suppresses the printing of the frequency table, which could be several pages long since each unique value of the SUV will be interpreted as a row of the table. The MEASURES option ensures that various measures of association, including Somers' D, are printed. Output 3.2 shows the results from this invocation of PROC FREQ.

Output 3.2

Statistic	Value	ASE
Gamma	0.7585	0.0498
Kendall's Tau-b	0.5374	0.0402
Stuart's Tau-c	0.6925	0.0561
Somers' D C|R	0.3889	0.0323
Somers' D R|C	0.7426	0.0523
Pearson Correlation	0.6187	0.0395
Spearman Correlation	0.6311	0.0466
Lambda Asymmetric C|R	0.6716	0.0634
Lambda Asymmetric R|C	0.0081	0.0180
Lambda Symmetric	0.2408	0.0318
Uncertainty Coefficient C|R	0.6252	0.0521
Uncertainty Coefficient R|C	0.1225	0.0104
Uncertainty Coefficient Symmetric	0.2049	0.0172
Sample Size = 181		

Two Somers' statistics are reported: C\|R uses the column variable as the predictor and the row variable as the gold standard, while R\|C uses the row variable as the predictor and the column variable as the gold standard. The relevant statistic here is R\|C, reported to be 0.7426 with a standard error of 0.0523. If the table was constructed so that the marker (the SUV) was the column variable and the gold standard was the row variable, then we would have used the Somers' D C\|R statistic.

Using the relationship

$$AUC = \frac{D+1}{2}$$

you can compute the area under the curve to be 0.8713, identical to the LOGISTIC output, which was reported with three significant digits. Also using the relationship

$$se(AUC) = \frac{se(D)}{2}$$

the standard error of AUC is computed to be 0.026. One can compute a confidence interval for AUC using asymptotic normal theory

$$AUC + 1.96 * se(AUC)$$

which turns out to be (0.8203, 0.9222). This is a more complete analysis than the one afforded by PROC LOGISTIC because you can judge the effects of sampling variability on the estimate of AUC.

3.4.3 The %ROC Macro

While PROC FREQ is sufficient for our purposes to compute the area under the curve and its standard error, now is a good time to introduce a SAS macro primarily used to compare the areas under several ROC curves; however, it can also handle a single ROC curve as a special case. The %ROC macro, currently in Version 1.5, is available for download at the following URL:

support.sas.com/samples_app/00/sample00520_6_roc.sas.txt

Following is the code that can be used for this example:

```
%roc(data=HeadNeck,var=SUV,response=gold,contrast=1,details=no,
     alpha=.05);
```

The macro variables used for invoking %ROC are

- Response: This is the gold standard, or the outcome. You need to create a data set variable that takes on values of 0 or 1 only; the macro can not recode the observed binary classes. This variable must be numeric; a character variable with only two distinct values (0 and 1) will pass the macro's error-checking facility but produce a PROC IML error, which is difficult to debug.

- Var specifies the marker.

- Contrast is used primarily to compare several curves, as we will see in later chapters. The value of Contrast is always 1 for a single ROC curve.

- Alpha pertains to confidence limit estimation.

- Details controls the amount of output printed.

Output 3.3 shows the results of this call. The area under the curve is 87.13%, the same as the concordance from PROC LOGISTC, and also the same as derived from Somers' D in PROC FREQ. The standard error is also the same as the standard error derived from Somers' D in PROC FREQ. Confidence intervals also match up, up to rounding error.

Output 3.3

```
The ROC Macro

         ROC Curve Areas and 95%
            Confidence Intervals
      ROC Area Std Error Confidence Limits

 SUV    0.8713   0.0263    0.8198    0.9228
```

Frequent ROC curve users should be familiar with this macro. If, on the other hand, you only occasionally need an AUC estimate and its standard error, then it may be easier to work with PROC FREQ.

3.5 Selecting an Optimal Threshold

In certain cases, although the original predictions are continuous, it is of interest to report a binary prediction. Recall the rain prediction example used at the beginning of this chapter. Suppose you have a model that provides a predicted probability of precipitation for a given day. Along with this predicted probability, you might want to display an icon that depicts rain to communicate the prediction in simpler terms. Most of us have seen such icons in television or newspaper weather forecasts. By including this icon we are essentially reporting a binary prediction (rain/no rain) based on a continuous predictor (predicted probability of rain). This is equivalent to choosing a threshold, which itself is equivalent to choosing an operating point on the ROC curve.

Sometimes external criteria may guide the choice of an operating point. In the absence of such criteria, you might choose a threshold that is optimal in some sense. There are two widely used ways of doing so:

1. Choose the threshold that will make the resulting binary prediction as close to a perfect predictor as possible.

2. Choose the threshold that will make the resulting binary prediction as far way from a noninformative predictor as possible.

To understand these methods better, remember that a perfect predictor has a single point on its ROC curve (namely, the upper left corner of the unit square) with 100% sensitivity and 100% specificity. In a similar vein, a noninformative marker's ROC curve lies along the diagonal of the unit square. The distance between a point on the ROC curve and these reference points is measured using the so-called *Euclidean method*. The Euclidean distance between points A and B with coordinates (x_1, y_1) and (x_2, y_2) is given as follows:

$$d_{AB} = \sqrt{(x_1 - x_2)^2 + (y_1 - y_2)^2}$$

To make this clear, the empirical ROC curve is plotted again in Figure 3.5 with two candidate optimal operating points, A and B. R denotes the reference point for the first method of choosing an optimal threshold. Similarly, P and Q denote the reference points for the second method. According to the first definition, you need to compute the length of line segments RA and RB and prefer A over B if RA is shorter than RB. Similarly, if the second method of choosing a threshold is adapted, then the lengths of AP and BQ need to be compared, with A preferred over B if AP is longer than BQ.

Figure 3.5 Choosing the Optimal Threshold Using the Empirical ROC Curve

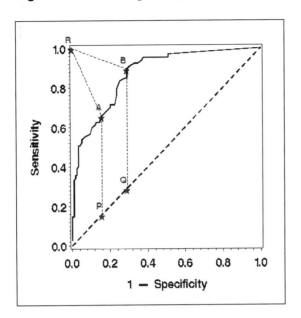

The OPTIMAL option of the %PLOTROC macro generates a data set called _OPTIMAL_, which computes the two distances explained here for each threshold. These two variables, called DIST_TO_PERFECT and DIST_TO_NONINF, can be used to choose an optimal operating point on the ROC curve.

The choice of method is mostly dictated by the field of application. In the context of medical diagnosis, optimal usually means *close to the ideal* and in engineering contexts it usually means *far from random*. In fact the vertical distance measure considered here is equal to Youden's index, a summary measure of accuracy that is popular in quality control, although it was first introduced in a medical context (Youden, 1950). *Youden's Index = how far from Random.*

Unfortunately, the two criteria of optimality produce different optimal operating points. This is analytically demonstrated by Perkins and Schisterman (2006). Most of the time, the two are close enough to allay any concerns about the choice of method. *Sens + Spe − 1*

3.6 The Binormal ROC Curve

So far our focus has been on the empirical curve, the one obtained by connecting the observed (sensitivity, 1-specificity) pairs. The empirical curve is widely used because it is easy to produce and it is robust (in the sense of being rank-based) to the changes in the marker distribution. But there are occasions when you might prefer a smooth curve. The most common way of smoothing an ROC curve is by using the binormal model.

The binormal model assumes that the distributions of the marker (or a monotone transformation) within the gold standard categories are normally distributed. For the PET data set example, we can assume that the SUVs of the patients with cancer follow a normal distribution, with mean μ_1 and variance σ_1^2, and SUVs of the patients with no cancer follow a normal distribution, with mean μ_0 and variance σ_0^2. Then, using the notation from the empirical ROC curves

$$F(t) = \Phi\left(\frac{\mu_0 - t}{\sigma_0}\right)$$

it follows that the threshold t can be written as a function of x as follows:

$$t = \mu_0 - \sigma_0 \Phi^{-1}(x)$$

Because a threshold t corresponds to the sensitivity $F(t)$, we can write the functional form of the ROC curve as follows:

$$G(t) = \Phi\left(\frac{\mu_1 - t}{\sigma_1}\right) = \Phi\left(\frac{\mu_1 - \mu_0 + \sigma_0 \Phi^{-1}(x)}{\sigma_1}\right) = \Phi\left(a + b\Phi^{-1}(x)\right)$$

where

$$a = \frac{\mu_1 - \mu_0}{\sigma_1}, \quad b = \frac{\sigma_0}{\sigma_1}$$

These two parameters, a and b, are often referred to as *binormal parameters*. Sometimes they are called *intercept* and *slope* because plotting a binormal curve on normal probability paper yields a straight line with intercept a and slope b. This practice is not common anymore, but the nomenclature continues.

If the marker values are not normally distributed but can be after a monotone transformation, then the binormal ROC curve is fitted to the transformed values. Nevertheless, the binormal curve fitted to the transformed values is still considered the ROC curve for the untransformed marker as well because the true ROC curve for the marker is invariant under monotone transformations, as shown in Section 3.3. Therefore, the existence of a monotone normalizing transformation is sufficient to justify using the binormal approach.

On the one hand, this is a relatively weak requirement, suggesting that most predictors can be thought of as binormal. On the other hand, discovering the functional form of h can be quite a task by itself. Note that the same $h(.)$ must transform the marker to normality for *both* gold standard positive *and* negative patients. This process can be challenging and is discussed in the next section.

The area under the curve for the binormal model also has a closed-form expression:

$$AUC = \Phi\left(\frac{a}{\sqrt{1 + b^2}}\right)$$

Estimation of the binormal model requires estimation of a and b, which in turn requires estimation of the mean and variances in each outcome group separately. This can simply be accomplished using PROC MEANS with a CLASS statement. But using PROC NLMIXED has many advantages, which the following example demonstrates. If you are not familiar with PROC NLMIXED, a relative newcomer to the SAS/STAT family, see the Appendix for an introduction covering its relevant capabilities.

```
proc nlmixed data=HeadNeck;
  parameters m1=0 m0=0 s1=1 s0=1;
  if gold=1 then m=m1;else if gold=0 then m=m0;
  if gold=1 then s=s1**2;else if gold=0 then s=s0**2;
  a=(m1-m0)/s1;
  b=s0/s1;
  model suv ~ normal(m,s);
  estimate 'a' a;
  estimate 'b' b;
  estimate 'AUC' probnorm(a/sqrt(1+b**2)));
run;
```

To adopt this code for your data set, replace the variables Gold and SUV with the variable names for the gold standard and the marker. The relevant portion of the NLMIXED output from this execution appears under Additional Estimates in Output 3.4.

Output 3.4

				Additional Estimates				
Label	Estimate	Standard Error	DF	t Value	Pr > |t|	Alpha	Lower	Upper
a	1.4086	0.1830	181	7.70	<.0001	0.05	1.0476	1.7696
b	0.6597	0.07196	181	9.17	<.0001	0.05	0.5177	0.8017
AUC	0.8802	0.02727	181	32.27	<.0001	0.05	0.8263	0.9340

You can see that the binormal parameters *a* and *b* are estimated to be 1.4086 and 0.6597. The corresponding ROC curve is represented by $\Phi(1.41 + 0.66\, \Phi^{-1}(x))$, where *x* takes on values from 0 to 1. That the ROC curve is defined for all points between 0 and 1 is a feature of the model. Although not all thresholds are observed, the binormal model *interpolates* and makes available a value of sensitivity for all possible values of one minus specificity.

The resulting AUC is 0.8802, with a standard error of 0.02727. You can form a 95% confidence interval using standard asymptotics: $AUC \pm 1.96 * se(AUC)$, which turns out to be (0.826, 0.934). Table 3.3 compares the results from the binormal model with those from the empirical analyses. Both the AUC estimates and the confidence intervals are very close to each other.

Table 3.3 Comparison of AUC Estimates from the Empirical and Binormal ROC Curves

Method	AUC	SE	Confidence Interval
Empirical	0.8713	0.0263	0.820–0.923
Binormal	0.8802	0.0273	0.826–0.934

From the standpoint of ROC curve analyses, the ESTIMATE statement is an important strength of PROC NLMIXED because it allows you to directly obtain estimates and, perhaps more importantly, standard errors for *a*, *b*, and *AUC*. Another advantage that will be apparent in subsequent chapters is that it is flexible enough to accommodate several different types of ROC curves.

The following is a graphical comparison of the empirical curve (solid line) and the binormal curve (dotted line) using the PET data set example. Early on, the binormal ROC curve is above the empirical one, suggesting that sensitivity is overestimated when compared with the empirical values. Later, the curves cross and the binormal curve starts to underestimate sensitivity. Most of the time the difference between the two curves is less than 10%. For most purposes, this would be considered a reasonable-fitting binormal ROC curve. Figure 3.6 shows a graph comparing the two different curves.

Figure 3.6 Comparison of Empirical and Binormal ROC Curves

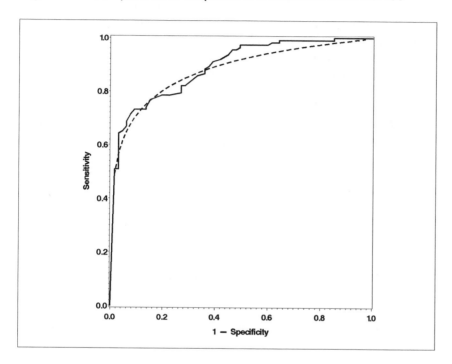

Figure 3.6 is generated by the following SAS statements. The key idea is in the DATA step, where the data set produced by the OUTROC= option in PROC LOGISTIC (called ROCData here) generates the sensitivities from the binormal model. ROCData already contains the sensitivity and one minus specificity pairs for the empirical curve. Another variable, called _BSENSIT_ here, is added using the functional form of the binormal model along with the parameters estimated with PROC NLMIXED (described previously). Then _BSENSIT_ and _SENSIT_ are plotted against _1MSPEC_, overlaid on the same graph. The solid line is the empirical curve (_SENSIT_) and the dotted line is the binormal curve (_BSENSIT_).

```
data ROCData;
  set ROCData;
  _bsensit_=probnorm(1.41+0.66*probit(_1mspec_));
run;


axis1 length=12cm order=0 to 1 by 0.2 label=(f=swissb h=2)
      value=(font=swissb h=2);
axis2 length=12cm order=0 to 1 by 0.2 label=(a=90 f=swissb h=2)
      value=(font=swissb h=2);

symbol1 v=none i=join w=3 l=1 c=black;
symbol2 v=none i=join w=3 l=3 c=black;
```

```
proc gplot data=ROCData;
   plot (_sensit_ _bsensit_)*_1mspec_ / haxis=axis1 vaxis=axis2
       overlay;
run;
quit;
```

3.7 Transformations to Binormality

The assumption of normality plays a crucial role in the binormal model. Statisticians who are familiar with the theory of *t*-tests and analysis of variance may dismiss this importance because of the widespread notion that large sample sizes overcome the effects of non-normality. This is generally true for procedures that compare the means of several groups (such as the *t*-test or the *F*-test), primarily because the sample means can be approximated by a normal distribution in large samples even if the individual observations are markedly non-normal.

In contrast, the ROC curve does not simply compare the means of diseased and non-diseased populations. In fact, as Section 3.3 shows, the ROC curve can be represented as $G(F^{-1}(x))$, where *x* is one minus the specificity, $G(.)$ is the cumulative density of the predictor in the diseased group, and $F(.)$ is the cumulative density in the non-diseased group. Therefore, accurate estimation of the ROC curve requires accurate estimation of **all points** on *F* and *G*. The binormal model assumes *F* and *G* to be the cumulative density function of the normal distribution, and the parameters of these two normal distributions are estimated from the data. If the data are markedly non-normal, estimates of *F* and *G* can be severely inaccurate at several points. This is not easily remedied by assembling larger samples.

For this reason, using the binormal model raises the question of whether the data are indeed normal and whether departures from normality can be fixed with a monotone transformation. Based on Section 3.3, you might think that all ROC curves are invariant. This is certainly not the case. All parametric ROC curves, and the binormal model in particular, are not invariant to monotone transformations.

Normalizing transformations is of great interest in least squares regression and analysis of variance, so there is a vast amount of literature about it. One of the most popular transformation methods is the standard Box-Cox power family implemented in PROC TRANSREG.

The Box-Cox power family of transformations can be defined as

$$y^{(\lambda)} = \begin{cases} \dfrac{y^{\lambda} - 1}{\lambda} & \lambda > 0 \\ \log(y) & \lambda = 0 \end{cases}$$

Each value of λ defines a transformation. For example, λ=1 uses the original values and λ=0 corresponds to a log-transform. This definition assumes *y* is non-negative to start with; if this is not the case, a constant can be added to shift the values. The idea is to choose λ based on the observed data. PROC TRANSREG uses a maximum likelihood estimate of λ along with the associated confidence interval. Because the choice of λ depends on the data, the recommendation is to round the resulting estimates to the nearest quarter so long as they are in the associated confidence interval.

This representation does not take into account the gold standard. The marker distributions that need normalization are the conditional ones (conditional on the gold standard, *F* and *G* in the foregoing notation). In contrast, the Box-Cox family attempts to normalize the marginal distribution of the marker values. You can redefine the family so that it normalizes the conditional distributions. The easiest way of doing so is recognizing that an analysis of variance model with the marker as the outcome and the gold standard as the covariate is one way of modeling the desired conditional distributions. For more information, see Sakia (1992).

In the PET example, you can estimate λ using the following call to PROC TRANSREG. The left-hand side of the model includes the marker value to be transformed and the right-hand side includes the gold standard, as explained previously, to ensure that conditional distributions are normalized. The PARAMETER= option specifies a constant to be added to all observations before λ is estimated. Because some SUVs are 0, an arbitrarily chosen small number (0.1) is added.

```
proc transreg data=HeadNeck;
   model boxcox(suv/parameter=0.1) = class(gold);
run;
```

Output 3.5 shows the results. The primary output is the (log) likelihood values for each λ. As explained in the legend, a less-than symbol (<) points to the best λ (that is, the one that maximizes the likelihood). Observations marked with an asterisk (*) form a confidence interval, and a plus sign (+) is a convenient choice for λ. According to the criteria for being convenient, the number must be round and contained in the confidence interval. In this example, λ=0.5 is identified as the best choice, implying a square root transformation.

Output 3.5

```
The TRANSREG Procedure

     Transformation Information
          for BoxCox(suv)

   Lambda        R-Square      Log Like

   -3.00           0.25        -1135.75
   -2.75           0.25        -1032.16
   -2.50           0.25         -930.05
   -2.25           0.25         -829.71
   -2.00           0.25         -731.55
   -1.75           0.25         -636.09
   -1.50           0.26         -544.03
   -1.25           0.26         -456.31
   -1.00           0.26         -374.17
   -0.75           0.27         -299.21
   -0.50           0.28         -233.40
   -0.25           0.31         -179.12
    0.00           0.34         -139.18
    0.25           0.38         -116.98  *
    0.50  +        0.42         -115.97  <
    0.75           0.43         -137.84
    1.00           0.42         -180.65
    1.25           0.39         -239.65
    1.50           0.35         -309.89
    1.75           0.30         -387.75
    2.00           0.26         -470.92
    2.25           0.23         -558.00
    2.50           0.20         -648.08
    2.75           0.17         -740.57
    3.00           0.14         -835.03

< - Best Lambda
* - Confidence Interval
+ - Convenient Lambda
```

You can create a data set containing the transformed values in PROC TRANSREG using the OUTPUT statement. These transformed values can then be used in a PROC NLMIXED call similar to the one in Section 3.6 to estimate the binormal parameters for the transformed variates. Although the steps for doing so aren't detailed here, you might want to try this exercise to gain practical skills.

3.8 Direct Estimation of the Binormal ROC Curve

Section 3.6 describes how to derive the functional form of a binormal ROC curve, starting with the basic assumption that the distribution of the marker is normal (with possibly different means and variances) in positive and negative groups. It then shows how PROC NLMIXED can be used to obtain estimates of the parameters of these two normal distributions. This method assumes that the observed data are normally distributed. However, as noted in Section 3.7, it is possible to transform the observed values to improve adherence to the normality assumption and then apply the approach outlined in Section 3.4 on the transformed values to obtain improved estimates of the ROC curve.

There is a simpler and more direct method to estimate the binormal ROC curve. Note that the binormal curve has only two parameters, a and b, and the ROC curve is given by

$$y = \Phi\left(a + b\Phi^{-1}(x)\right)$$

where y is the sensitivity and x is one minus the specificity. You can first plot the observed ROC points, as if for an empirical ROC curve, but instead of connecting them, you can fit a curve with the form above and find a and b such that the resulting fit is best. This process is further facilitated by the observation that

$$\Phi^{-1}(y) = a + b\Phi^{-1}(x)$$

Therefore, if the ROC points are first transformed to the probit scale, then you only need to fit a line, which can be done, almost trivially, by using least squares. There are many SAS/STAT procedures available for this purpose, but PROC REG offers the simplest interface. Suppose ROCData is a data set created by the OUTROC= option of PROC LOGISTIC. Then you can estimate the binormal parameters by using the following code:

```
data ROCData;
   set ROCData;
   _y=probit(_sensit_);
   _x=probit(_1mspec_);
run;

proc reg data=ROCData;
   model _y=_x;
run;
quit;
```

This method leads to estimates of 1.71 and 0.85 for the binormal parameters a and b, respectively, in the head and neck cancer data, compared with 1.31 and 0.66 when the binormal parameters were estimated by modeling the marker values directly using PROC NLMIXED. Figure 3.7 shows the two different binormal fits to the empirical curve (solid line). The dashed line is the PROC NLMIXED estimate and the dotted line is the PROC REG estimate.

Figure 3.7 Estimates of the ROC Curve Obtained from NLMIXED, REG, and LOGISTIC Procedures

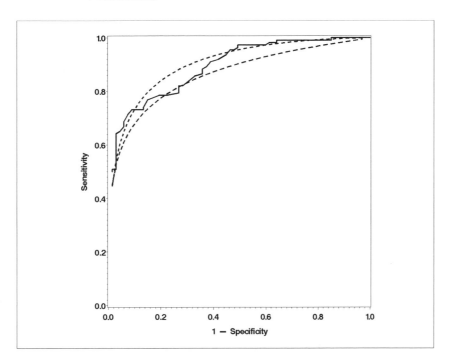

Although the direct estimation of the ROC curve using least squares is attractive because of its simplicity, it is not without problems, especially the fact that the points on the ROC curve are in fact points from a cumulative distribution function and are not independent. So PROC REG standard errors are likely to be underestimates of the true standard errors. Therefore, you should not use least square estimates for inference (such as reporting confidence intervals) but as point estimates of the binormal parameters. Since the binormal curve is a function of the binormal parameters, direct estimation of the ROC curve using the least squares estimates is correct, but the resulting standard errors and *p*-values are incorrect.

3.9 Bootstrap Confidence Intervals for the Area under the Curve

Bootstrap is a modern method that originally started for assessing the variability in point estimates. It has since expanded to perform hypothesis tests, validation of predictive models, model selection, and many other statistical tasks.

In this book, bootstrap methods are used for

- Non-parametric confidence intervals for the area under the ROC curve
- Hypothesis tests that compare two or more ROC curves with respect to their AUCs
- Correcting for the optimism in the evaluation of multivariable prediction models, such as logistic regression, or data mining predictions, such as classification trees.

Very briefly, a bootstrap sample is obtained by sampling from the observed data with replacement. The following steps describe the process in pseudo-code:

1. Create *B* data sets, each random samples *with* replacement from the original data.

2. For *i*=1,…,*B*, analyze data set *i* exactly as the original data set would have been analyzed (for example, compute the empirical AUC). Assign the results (possibly several values) to *R*(*i*).

3. *R*(1), … , *R*(*B*) is an approximation of the sampling distribution of *R*.

The approximation works better for large *B*. In most applications, *B* is between 200 and 2,000. Statistics estimated from the approximate sampling distribution serve as the desired estimates. For example, the standard deviation of the *B* numbers *R*(*i*) is approximately the standard error of *R*. Further review of the bootstrap method is beyond the scope of this book. Consult Efron and Tibshirani (1993) for more information.

The macro %BOOT1AUC, which is available from the book's companion Web site at support.sas.com/gonen, performs a bootstrap analysis for the area under the curve of a single ROC curve. An example call is as follows:

```
%boot1auc(data=HeadNeck,marker=suv,gold=ngold);
```

although there are two other possible inputs: BOOT specifies the number of bootstrap samples (default of 1,000) and ALPHA specifies the error level for the confidence intervals. Output 3.6 shows the results.

Output 3.6

```
Bootstrap Analysis of the AUC for suv

    AUC      StdErr
  _____

0.871523   0.026945


95% Confidence Interval

  LowerLimit      UpperLimit
  _____

    0.813384        0.920954
```

If you want to further develop your own bootstrap applications, PROC SURVEYSELECT provides the easiest way to create a bootstrap sample with SAS:

```
proc surveyselect data=HeadNeck method=urs n=181 out=_sample outhits
      rep=100 noprint;
run;
```

This call to PROC SURVEYSELECT generates 100 samples (as specified with the REP= option) and each sample has 181 elements (as specified by the N= option). In general, each bootstrap sample has the same size as the original data set, so the N= option typically equals the sample size. METHOD=URS allows unrestricted random sampling and corresponds to a sampling schema that gives equal probability to each record in the data set and performs the sampling with replacement. The OUT= option specifies the data set in which the bootstrap samples are saved. Finally, the OUTHITS= option ensures that a record that was selected more than once (which is possible in sampling with replacement) appears as many times as it was selected in the output data set. The default (when the OUTHITS= option is not specified) represents that record with a single observation, along with a variable that indicates the number of hits (i.e., the number of times it was selected).

An important feature of the data set created by PROC SURVEYSELECT is that it contains a variable called Replicate that identifies the bootstrap sample. Because bootstrap samples are analyzed separately, you can use the BY processing available in most SAS/STAT procedures. For example, to compute the AUC for each bootstrap sample, you can call PROC LOGISTIC using a BY REPLICATE statement and create an output data set using the ODS OUTPUT statement. This output data set will also include the REPLICATE column along with the AUC index computed for each bootstrap sample. Therefore, developing SAS code for the bootstrap method has become substantially easier with PROC SURVEYSELECT. In addition, because BY processing is much faster than competing ways of performing the same task, such as using macro loops, this strategy leads to computationally efficient programs. This feature is critical because bootstrap methods can require intensive computer resources.

Comparison and Covariate Adjustment of ROC Curves

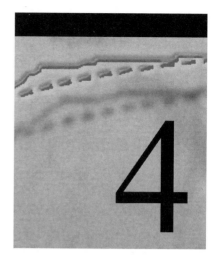

4.1 Introduction

In many instances there will be a set of competing predictions to choose from and hence it is often of interest to compare predictions. For example, how does an expensive invasive procedure or radiological scan compare with a simple clinical exam to diagnose a medical condition? Or, is it worth it to obtain at substantial cost detailed credit reports while one already has a simple financial summary of the debtors. Does the incremental increase in accuracy justify the additional cost? The term *cost* here is used loosely; for example, the additional burden on a patient of performing an additional diagnostic scan is considered a cost. Of course, there are traditional costs to consider, too: The monetary cost of obtaining the additional scan needs to be taken into account as well. Before deciding whether these costs are justified, you need to evaluate the gains in predictive accuracy.

A common and simple way to compare competing predictors is to construct an ROC curve for each one. Visual comparisons of these ROC curves usually reveal several useful features. In addition, two ROC curves can be compared for statistically significant differences. This chapter discusses non-parametric and parametric methods of statistically comparing two or more ROC curves.

4.2 An Example from Prostate Cancer Prognosis

Chapter 3 used an example from cancer diagnosis and monitoring to introduce ROC curves. This chapter uses an example from cancer prognosis to study the methods of comparing two ROC curves. This time, the data are obtained on prostate cancer patients with bone metastases. Traditionally, these patients undergo a *bone scan*, a radiographic exam that reveals the abnormalities in the bone. You can compute a summary measure, called a *bone scan index* (*BSI*), which indicates the extent of bone disease in patients. A competing marker is the SUV from a PET scan that was used in Chapter 3 to identify head and neck cancer. The prostate data set contains measurements of BSI and SUV for 51 patients. The outcome of interest is one-year survival—that is, whether patients are alive one year after their BSI and SUV are measured. Each patient was followed at least one year and the outcome recorded. Forty patients were dead, and 11 were alive at the one-year mark.

This example is used throughout the chapter to compare the ROC curve of SUV with the ROC curve of BSI. But before further discussion, we need to distinguish between paired and unpaired mechanisms of data collection, an important issue associated with most data sets involving ROC curve comparisons.

4.3 Paired versus Unpaired Comparisons

When dealing with any statistical comparison, you need to distinguish unpaired and paired data. Here, *unpaired* refers to when data in the two groups to be compared are obtained from different sampling units and *paired* refers to when each sampling unit contributed exactly a pair of observations: one to each group.

Consider the example of a randomized clinical trial that evaluates screening mammography. Women are randomized to either the standard of care, which is periodic clinical exams, or to mammography, which is mammography in addition to the standard of care. The resulting comparison, whether performed using ROC curves or not, is unpaired. Each woman in the study is either in one study arm or the other.

Now, consider another study that evaluates PET in imaging tumors before surgery. Suppose that there is a standard imaging tool for this purpose, likely computed tomography (CT) for most cases. One possibility is randomizing the patients to receive either a CT or a PET, creating an unpaired comparison. However, a statistically more efficient way of comparing the two is arranging for the patients to undergo both scans, creating a paired data situation.

In traditional statistics, most comparisons involve unpaired data as opposed to paired data. For example, the two-sample (unpaired) *t*-test is probably the most widely applied statistical test. The reason for this is not increased efficiency. In fact, for a fixed sample size, paired techniques are usually more powerful than unpaired ones. It is, however, more difficult, or sometimes impossible, to implement a paired design. Randomized trials of many drugs and interventions use unpaired designs because exposing the subject to multiple treatments or interventions is often difficult, if not impossible. Imagine a trial comparing surgical treatment against watchful waiting, a common option, for early stage indolent tumors such as some cancers of the prostate. It is not possible for patients to undergo both treatments.

Paired comparisons have made their mark as the design of choice in some fields, including comparison of predictive markers. Many of the markers under consideration can be applied to each sampling unit without too much burden, cost, or undue consequence. Thus, this chapter focuses on comparing ROC curves with paired data, although references to unpaired comparisons will be noted.

4.4   Comparing the Areas under the Empirical ROC Curves

In addition to the distinction between paired and unpaired data, you also need to decide what comparing the ROC curves means. It could refer to comparing the curves in their entirety, or it could refer to comparing certain summary measures, such as the area under the curves.

When dealing with empirical ROC curves, the usual choice is comparing the summary measures. The alternative, comparing entire curves, is a difficult task without the help of a model. The caveat, of course, is that two predictors may have the same AUC but very different ROC curves. For this reason, a graphical display of the ROC curves under comparison should always accompany the statistical test comparing the AUCs.

4.4.1   Comparisons with Paired Data

You can organize the paired data in two ways. The most intuitive way is to have one row for each subject and one column for each marker in addition to a column for the outcome. An equivalent way is to have one row for each subject-marker combination, resulting in two rows per subject. The outcome variable will be duplicated in these two rows and an additional column containing the subject identifier will be added to keep track of the pairing. These two data structures are known as *wide* and *long*, respectively, with obvious reference to the shape of the data matrix. One can be converted to the other with simple DATA step instructions.

The %ROC macro that was used in Chapter 3 to obtain estimates of the standard error of the AUC can also be used to compare two ROC curves *only if the data are paired*. The methodology implemented in this macro is essentially non-parametric. Each AUC is estimated using the concordance method described in the previous chapter. The concordance statistic is a member of the class of U-statistics. The variance of the AUCs as well as the covariance between them can be computed using the general principles of the theory of U-statistics. See DeLong et al. (1988) for more information.

The %ROC macro expects the data in the wide form. The following macro call performs the needed comparison:

```
%roc(data=bone, var=bsi suv1, response=gold, contrast=1 -1);
```

The VAR macro variable specifies the names of the variables containing the two marker values. The CONTRAST macro variable expects input similar to the CONTRAST statement in PROC GLM, so 1 -1 refers to the difference of the two variables stated in the VAR macro variable.

Output 4.1 shows the results of this call. The AUC estimates for each of the ROC curves, along with their standard errors and (marginal) confidence intervals, appear first. BSI has an AUC of 0.7795, which compares favorably with that of SUV at 0.6773. The standard errors are quite large, however: 0.0771 for BSI and 0.0836 for SUV, resulting in wide confidence intervals. The AUC for BSI has a 95% confidence interval that ranges from 0.6284 to 0.9307. The confidence interval for SUV is between 0.5134 and 0.8412. It seems that SUV is barely distinguishable from a random predictor (since the lower confidence limit barely exceeds 0.50).

Output 4.1

```
The ROC Macro


          ROC Curve Areas and 95%
             Confidence Intervals
      ROC Area Std Error Confidence Limits

BSI    0.7795   0.0771    0.6284      0.9307
suv1   0.6773   0.0836    0.5134      0.8412



 Contrast Coefficients
          BSI       suv1


Row1       1        -1


      Tests and 95% Confidence Intervals for Contrast Rows
      Estimate  Std Error Confidence Limits   Chi-square Pr > ChiSq


Row1  0.1023    0.0954   -0.0847     0.2893      1.1491  0.2837


     Contrast Test Results
Chi-Square    DF      Pr > ChiSq

  1.1491      1        0.2837
```

Contrast information appears next. This can be used to verify that the results are for the intended comparison and to infer the direction of the difference. In this example, a positive difference in AUC curves points to BSI being the better test. The next piece of output compares the two ROC curves. The difference in the two AUCs is 0.1023, as reported under the Estimate column. This could be obtained directly as the difference between 0.7795 and 0.6773, the AUCs for BSI and SUV reported in the first part of the output. However, you can not compute the standard error of the difference from the information about the individual areas because the data are paired. Based on this standard error, the difference between the two AUCs is not significant ($p=0.2837$) and the confidence interval of (-0.0847, 0.2893) spans 0.

It appears that the test statistic and the p-value are reported twice. Had the contrast involved more than one row, then the output on contrast rows would have included one line for each row in the contrast. The part labeled "Contrast Test Results" would still report a single p-value with $r-1$ degrees of freedom, where r is the number of rows. Therefore, only in a contrast with a single row does information in the "Contrast Rows" section and the "Contrast Test" section coincide.

As discussed here, it is essential to present a visual display of the ROC curves along with the comparison of the AUCs. Figure 4.1 is generated by the %PLOTROC macro introduced in Chapter 3. The syntax modification is similar to the one in the %ROC macro (that is, multiple markers are specified in the VAR macro variable, separated by blanks).

```
%plotroc (bone bsi,suv,gold,anno=4,tlist1=1.08 2.29, tlist2=4.8 5.8);
```

Figure 4.1 ROC Curves for BSI (Solid) and SUV (Dashed) Overlaid

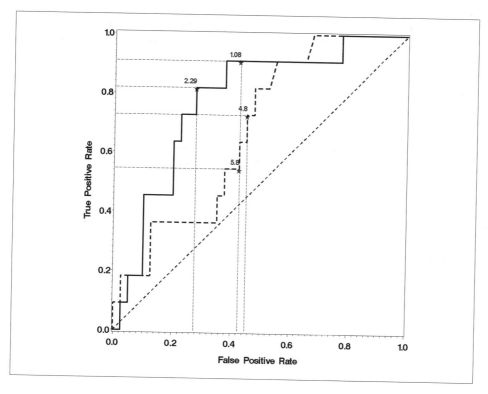

Figure 4.1 shows that the BSI has a higher sensitivity than the SUV for most values of the specificity. In the range where sensitivity is 50% through 80%, the sensitivity of BSI is substantially higher than that of SUV. A threshold of 2.29 for the BSI achieves a sensitivity of more than 80% and a specificity of more than 70%. None of the operating points for the SUV offers comparable accuracy. Despite this, keep in mind that the AUC estimates displayed high variability and the difference was not statistically significant.

4.4.2 Comparisons with Unpaired Data

In theory, the %ROC macro can only compare the AUCs of two predictors if the data are paired. Because unpaired designs are so uncommon in the field of ROC curves, no macro performs the desired comparison. Fortunately, unpaired comparisons are easy to be performed by hand or by a few lines of DATA step code: If AUC_1 and AUC_2 are the two AUC estimates with associated standard errors s_1 and s_2, then the test statistic for the unpaired comparison is as follows:

$$T = \frac{\left(AUC_1 - AUC_2\right)^2}{s_1^2 + s_2^2}$$

In a large sample, T will have a chi-square distribution with one degree of freedom under the null hypothesis (that is, when the two AUCs are indeed equal).

Note that AUC_1 and s_1 are reported when the %ROC macro is executed for the first predictor only, as explained in Chapter 3. Similarly, AUC_2 and s_2 are estimated from a second %ROC call. Once T is computed, the two-sided p-value can be found using the following line in a DATA step:

```
p_value=1-probchi(t_unpaired,1);
```

If the prostate cancer study offered either a PET or a BSI (but not both) so that the resulting data structure was unpaired, and if the AUC estimates and their standard errors were the same as described previously (0.7795 and 0.6773 for the AUCs, 0.0771 and 0.0836 for the standard errors), then T would be 0.8095 and the p-value would be 0.3683. The increase in p-value using the unpaired design can be explained by the increased efficiency of the paired design, as briefly discussed in Section 4.3.

4.5 Comparing the Binormal ROC Curves

Chapter 3 shows that you can achieve a model-based estimate of the ROC curve by assuming that the distributions of the marker for the diseased and non-diseased groups have normal distributions with possibly different parameters. Specifically, if diseased patients' marker values follow a normal distribution with mean μ_1 and variance σ_1^2 and if non-diseased patients' marker values follow a normal distribution with mean μ_0 and variance σ_0^2, then the ROC curve can functionally be written as

$$\Phi\left(a+b\Phi^{-1}(x)\right)$$

with the AUC given by

$$\Phi\left(\frac{a}{\sqrt{1+b^2}}\right)$$

where the binormal parameters a and b are given by

$$a=\frac{\mu_1-\mu_0}{\sigma_1}, \quad b=\frac{\sigma_0}{\sigma_1}$$

Next, we discuss paired data in detail; comments about unpaired data appear at the end of this section.

4.5.1 Comparisons Based on the Binormal Model with Paired Data

The paired nature of the data requires some adjustment to the model before you can proceed. Let X_{ij} denote the marker value for the ith patient and the jth marker. Then X_{ij} (or some transformation of it) is assumed to have the following normal distribution if the patient has a positive outcome:

$$X_{ij} \sim N\left(\mu_{1j}+u_i,\sigma_{1j}^2\right)$$

If the patient has a negative outcome, then X_{ij} would be distributed as

$$X_{ij} \sim N\left(\mu_{0j} + u_i, \sigma_{0j}^2\right)$$

The u_i is normally distributed as well:

$$u_i \sim N\left(0, \sigma_r^2\right)$$

with the key assumption that u_i and $u_{i'}$ are independent for i≠i'.

This model has four fundamental parameters for each marker: μ_{1i} and μ_{0i} are the means of the outcome positive and outcome negative patients for marker j, and σ_{1i} and σ_{0i} are the two standard deviations. In this respect, it is a standard binormal model for each marker separately. To take into account the correlation between the two marker measurements on the same subject, a random effect u is introduced. Because u_i is a subject-specific effect, the two marker values that are measured on the same patient will have the same u_i. This commonality induces a correlation on the two measurements (one for each marker) taken on the same subject. Because u_i and $u_{i'}$ are independent for i≠i', there is no correlation between marker values that were not measured on the same subject. Use of random effects to induce correlations on selected pairs of subjects is a standard technique in most settings.

The following code segment first runs PROC TRANSREG to identify a transformation to normality. See Section 3.7 for a detailed explanation of the role of this step in fitting binormal models.

```
proc transreg data=bone;
   model boxcox(suv1/parameter=0.1 geometricmean)=class(gold);
   output out=trsuv;
run;

proc transreg data=bone;
   model boxcox(bsi/parameter=0.1 geometricmean)=class(gold);
   output out=trbsi;
run;
```

The Box-Cox model identifies the logarithmic transformation for the BSI and the square-root transformation for the SUV as the best possible transformations to normality (detailed PROC TRANSREG output not shown). The output data sets TRSUV and TRBSI are created by the OUTPUT statement in PROC TRANSREG and include the transformed covariates, TBSI and TSUV. These two data sets are combined to form Long, a data set in long form. Then PROC NLMIXED finds the maximum likelihood estimates for the random-effects binormal model, as specified previously.

```
data long;
   set trbsi(keep=_name_ gold tbsi rename=(tbsi=result _name_=subid)
         in=in1)
      trsuv(keep=_name_ gold tsuv rename=(tsuv=result _name_=subid)
         in=in2);
   if in1 then marker=0; else if in2 then marker=1;
run;
```

```
proc nlmixed data=long method=firo technique=quanew;
    parms sr=1 s11=3 s10=0.3 s01=2 s00=0.2 m11=7 m10=1 m01=2 m00=0;
    bounds sr s11 s10 s01 s00>0;
    if gold=1 and marker=0 then do;
                m=m10+u;s=s10**2;end;
    if gold=1 and marker=1 then do;
                m=m11+u;s=s11**2;end;
    if gold=0 and marker=0 then do;
                m=m00+u;s=s00**2;end;
    if gold=0 and marker=1 then do;
                m=m01+u;s=s01**2;end;
    a1=(m11-m01)/s11;
    b1=s01/s11;
    a0=(m10-m00)/s10;
    b0=s00/s10;
    auc1=probnorm(a1/sqrt(1+b1**2));
    auc0=probnorm(a0/sqrt(1+b0**2));
    model result ~ normal(m,s);
    random u ~ normal(0,sr) subject=subid;
    estimate 'a1' a1;
    estimate 'b1' b1;
    estimate 'AUC1' auc1;
    estimate 'a0' a0;
    estimate 'b0' b0;
    estimate 'AUC0' auc0;
    estimate 'AUC1-AUC0' auc1-auc0;
    contrast 'AUC1-AUC0' auc1-auc0;
    contrast 'Equality of ROC curves' a1-a0, b1-b0;
run;
```

The fundamental description of the model in PROC NLMIXED is similar to the one in Section 3.6, with two exceptions of the random effect u. The RANDOM statement declares u as the random effect. For those familiar with PROC MIXED, the RANDOM statement has similar functionality but different syntax. In PROC MIXED, you don't need to specify the distribution of the random effect since normal random effects are the only available option. In PROC NLMIXED, you must specify the distribution of the random effects since any of the built-in distributions that PROC NLMIXED recognizes is an option. Another difference is that, in PROC MIXED, specifying the RANDOM statement ensures that the random effect is added to the model; in PROC NLMIXED, you must add the term u to the mean function in programming statements. See the appendix for a brief introduction to the syntax and capabilities of PROC NLMIXED.

In addition to the presence of the random effect, this call to PROC NLMIXED is different from the ones in Chapter 3 in a few other ways. For example, the CONTRAST statement provides simultaneous testing of parameters. It is helpful to think of it as multiple ESTIMATE statements executed simultaneously. To make this point, note that

```
estimate 'AUC1-AUC0' auc1-auc0;
```

and

```
contrast 'AUC1-AUC0' auc1-auc0;
```

perform the same test (that is, the equality of the AUCs). You would normally do this by using the ESTIMATE statement; including the CONTRAST statement for this purpose is for demonstration only.

The more important use of the CONTRAST statement here is to test the equality of the two ROC curves in their entirety. Note that two binormal ROC curves are identical only if they have the same binormal parameters. Therefore, comparing a and b for two or more binormal curves offers one way of testing whether the two curves are identical. This is accomplished by the following CONTRAST statement:

```
contrast 'Equality of ROC curves' a1-a0, b1-b0;
```

which tests the null hypothesis

$$H_0 : a_0 = a_1 \quad \& \quad b_0 = b_1$$

against the alternative that at least one of the binormal parameters is different. Note that it is not possible to express this using an ESTIMATE statement.

Output 4.2 shows the relevant portion of the output.

Output 4.2

Parameter Estimates

Parameter	Estimate	Standard Error	DF	t Value	Pr > |t|	Alpha	Lower	Upper
sr	2.5830	42.7645	101	0.06	0.9520	0.05	-82.2504	87.4163
s11	1.2701	16.8601	101	0.08	0.9401	0.05	-32.1758	34.7159
s10	1.3251	16.1371	101	0.08	0.9347	0.05	-30.6866	33.3368
s01	2.8855	7.4207	101	0.39	0.6982	0.05	-11.8353	17.6062
s00	1.8913	11.3087	101	0.17	0.8675	0.05	-20.5422	24.3248
m11	4.8146	0.6176	101	7.80	<.0001	0.05	3.5894	6.0398
m10	2.3499	0.6280	101	3.74	0.0003	0.05	1.1040	3.5958
m01	2.3526	0.5222	101	4.50	<.0001	0.05	1.3166	3.3885
m00	-0.1234	0.3924	101	-0.31	0.7538	0.05	-0.9019	0.6551

Contrasts

Label	Num DF	Den DF	F Value	Pr > F
AUC1-AUC0	1	101	0.01	0.9369
Equality of ROC curves	2	101	0.00	0.9963

Additional Estimates

Label	Estimate	Standard Error	DF	t Value	Pr > |t|	Alpha	Lower	Upper
a1	1.9385	25.7415	101	0.08	0.9401	0.05	-49.1258	53.0028
b1	2.2719	24.3351	101	0.09	0.9258	0.05	-46.0023	50.5462
AUC1	0.7826	0.9919	101	0.79	0.4320	0.05	-1.1850	2.7502
a0	1.8665	22.7369	101	0.08	0.9347	0.05	-43.2374	46.9704
b0	1.4273	8.8715	101	0.16	0.8725	0.05	-16.1713	19.0259
AUC0	0.8579	1.9318	101	0.44	0.6579	0.05	-2.9742	4.690
AUC1-AUC0	-0.07534	0.9492	101	-0.08	0.9369	0.05	-1.9583	1.8076

The Parameter Estimates section of the output presents estimates of the model parameters and is of little use here. The more interesting part for ROC analysis is the Additional Estimates section, which provides information on the results of the ESTIMATE statement, including binormal parameters of BSI (a_0 and b_0) and SUV (a_1 and b_1). This section also reports the implied AUC (AUC_1 and AUC_0) for each marker and the difference between the two AUCs. We see that the AUCs are estimated to be 0.7825 for the SUV and 0.8579 for the BSI. Their difference is 0.0753 and is not significant, with $p=0.9369$.

The two contrasts are reported in the Contrasts section. The first one is the difference between the two AUCs, repeated here only to highlight the similarities between the ESTIMATE and CONTRAST statements. This contrast has the same p-value as the ESTIMATE statement (as it should) but uses the F-statistics rather than t-statistics. If you are familiar with analysis of variance, you will remember that an F-statistic with a single numerator degree of freedom is identical to the square of the corresponding t-statistic. The same principle applies here, although it is hard to see from this output since the test statistics have very small absolute values. Nevertheless, note that $t=-0.08$, the square of which is 0.01 when rounded to two decimal places.

The contrast of real interest is the simultaneous test of the equality of the two binormal parameters. The results strongly suggest that there is no evidence against the equal ROC curves hypotheses ($p=0.99$).

4.5.2 Comparisons Based on the Binormal Model with Unpaired Data

As explained in Section 4.5.1, the random effect u was introduced to the binormal model only to account for the within-subject correlation. In unpaired designs, the within-subject correlation is 0 by definition. This suggests a simple way to modify the code from the previous section: Remove u. This amounts to removing the RANDOM statement and cleaning up the way m is defined in the programming statements.

4.6 Discrepancy between Binormal and Empirical ROC Curves

It is not uncommon for the binormal model and the empirical model to reach different conclusions. The difference between them is similar to the difference between rank tests and t-tests. If the assumptions underlying the binormal model are true, then the binormal model has more power, which might explain the significant result. On the other hand, if the model assumptions are not true, then the Type I error may be inflated, which would appear as increased false positive results. Although it is not possible to conclude which one is the driving force, certain features of the PROC NLMIXED output, along with exploratory graphical analyses like the one in Figure 4.2, remain the best way of checking the assumptions of binormality.

Figure 4.2 overlays the empirical and the binormal ROC curves, the latter as estimated by PROC NLMIXED. The solid lines indicate the BSI and the dashed ones indicate the SUV. The step function signifies an empirical ROC curve, while the smooth one follows from the binormal model. Both binormal curves are poor fits to the empirical ones. This also explains the discrepancy between the AUC estimates. For BSI, the empirical AUC is 78% while the binormal AUC is 85%, and for SUV the corresponding estimates are 68% and 78%.

Figure 4.2 Comparison of ROC Curves for Prostate Cancer Data

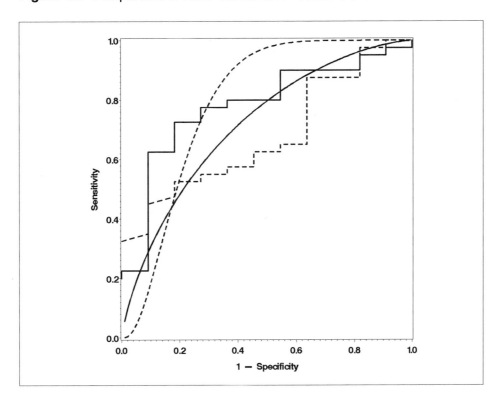

Another sign of poor fit is the unusually large estimates of variability for the model parameters. The estimated values of a and b look normal, but the standard errors are 10 to 20 times larger than model parameters. Although this can happen when estimates are near 0, the estimates in this model are not close to 0.

Be aware that despite their enormous popularity, the binormal models make some strong assumptions. The binormal ROC curve is no longer rank-based; every single observation contributes to the estimation of the binormal slope and intercept. Deviations from normality may have undue influences on the estimated curve and consequences on statistical inference. Generally, there is little reason to use the binormal model to compare ROC curves since methods based on the empirical curve are efficient and implemented in SAS.

You should, however, keep the binormal model in your toolbox. When continuous covariates might affect the predictive accuracy of the marker, empirical ROC methods can no longer be used. You must use model-based techniques, and the binormal model (along with the Lehmann family, covered in subsequent chapters) provides a comprehensive framework from which to draw inferences.

4.7 Bootstrap Confidence Intervals for the Difference in the Area under the Empirical ROC Curve

The bootstrap idea developed in Chapter 3 can be extended to get a confidence interval for the difference in AUCs as well as a *p*-value for testing whether the difference is significant. The strategy is similar:

- Generate B bootstrap samples.

- For each sample i, compute the AUC of the two ROC curves, $AUC_0^{(i)}$ and $AUC_1^{(i)}$ and compute the difference $\Delta^{(i)} = AUC_0^{(i)} - AUC_1^{(i)}$.

- The B numbers, $\Delta^{(1)}$ through $\Delta^{(B)}$, approximate the sampling distribution of the difference between the two AUCs.

This can be accomplished by using the %BOOT2AUC macro, which is available from the book's companion Web site at support.sas.com/gonen. The %BOOT2AUC macro is similar to the %BOOT1AUC macro introduced in Section 3.9. The only difference is that it allows for two variable names for the VAR macro variable. For example, the following call requests a comparison of SUV and BSI:

```
%boot2auc(HeadNeck,suv bsi,ngold);
```

The results appear in Output 4.3.

Output 4.3

```
     AUC1        SE      AUC2        SE  Difference         SE

 0.761397  0.080334  0.674859  0.080334    0.086573  0.123361


 95% Confidence Interval for Difference

 LowerLimit   UpperLimit
 ──────────────────────────
  -0.24815     0.280488


 H0: Difference=0
 p=0.341
```

The first part of the output reports the point estimates and standard errors for the individual AUCs as well as their difference. The AUCs and their standard errors are presented and labeled in the order specified in the macro call. The BSI has a bootstrap-estimated AUC of 0.761, and the SUV has 0.675. The difference is 0.087, considerable clinically but not statistically significant with $p=0.341$ and a confidence interval of $(-0.248, 0.280)$, which includes 0.

4.8 Covariate Adjustment for ROC Curves

In some contexts, covariates influence the accuracy of predictions. In this case, you need to adjust the ROC curve for these covariates. For example, in the context of weather forecasting, the same model that produces the forecasts may have varying accuracy according to the altitude of the geographical location for which the forecasts are produced. In the field of medical diagnosis, certain patient aspects, such as previous exposures to therapy, may influence the ability of diagnostic tools to accurately identify their current medical status. Finally, in the field of credit scoring, the same scoring method may have variable performance in different countries, perhaps due to different legal requirements on how the financial results are reported.

It is important to distinguish a covariate that is a predictor of the outcome as opposed to a covariate that is a risk factor for the outcome. To make this distinction concrete, consider an example from the field of medical diagnostics. Lymphoma is the cancer of lymph nodes. A malignant node is typically enlarged, so the size of a lymph node can be a good predictor of lymphoma. Malignancy, however, is not the only reason that lymph nodes grow. Lymph nodes can be enlarged in patients fighting an infection. Therefore, the size of the lymph node is likely to be a good predictor in patients who did not have a recent infection but a poor predictor if the patient has recently had an infection. On the other hand, whether someone had a recent infection is not a good predictor of lymphoma. Therefore, recent infection is a covariate that may affect the ROC curve but not a covariate that may be used to predict the presence of cancer, neither by itself nor in combination with another predictor.

In contrast, a family history of lymphoma (or perhaps any cancer) makes it more likely for a patient to have cancer because some cancers are genetically inherited. This may be factored into the diagnosis, formally or informally. Whether a patient has a family history of cancer usually has no bearing on whether an enlarged lymph node is a good predictor of cancer. Therefore, a family history of cancer is not a candidate for adjusting ROC curves.

In its simplest form, for a categorical covariate, adjustment implies estimating separate ROC curves for each value of the covariate. With continuous covariates, you can envision an infinite number of ROC curves, one for each possible covariate value. This requires the use of a model that postulates a relationship between the covariate and the parameters of the ROC curves. It should be no surprise that the primary such model is regression. The next section investigates how the binormal model studied in Chapter 3 can be formulated as a regression model.

4.9 Regression Model for the Binormal ROC Curve

The easiest regression model does not contain any covariates. It is not a useful model in practice, but it helps to frame the concepts. In the context of ROC curves, if T denotes the marker and D denotes the disease status (gold standard), then you can write

$$T = \alpha_0 + \alpha_1 D + e^{2D\beta}\varepsilon$$

where ε has a normal distribution with mean 0 and variance σ_0^2. After some algebra, you can show that

$$a = \frac{\alpha_1}{e^{\beta}\sigma_0^2}, \quad b = e^{\beta}$$

Remember that the coefficient of the variable D represents the difference of the means of the marker for $D=1$ and $D=0$. This holds true for any linear model and it will be used in subsequent chapters to extract the information relating to the ROC curve from more complicated models.

This model cannot be fit using the standard linear model procedures in SAS/STAT software, such as GLM or REG. It can be fit, however, with PROC MIXED and, as we saw in Chapter 3, with PROC NLMIXED. PROC MIXED syntax is easier because it follows the general outline of most of the SAS/STAT procedures. On the other hand, PROC NLMIXED can be generalized to the case of ordinal markers as well, as we will see in Chapter 5, so there is a value in adopting PROC NLMIXED as the primary choice of SAS/STAT procedure for modeling ROC curves.

Now we can model the covariates jointly. Consider a binary marker W first:

$$T = \alpha_0 + \alpha_1 D + \alpha_2 W + e^{2\beta}\varepsilon$$

which implies

$$T = \begin{cases} \alpha_0 + \alpha_1 D + e^{2D\beta}\varepsilon, & W = 0 \\ \alpha_0 + \alpha_2 + \alpha_1 D + e^{2D\beta}\varepsilon, & W = 1 \end{cases}$$

These two equations represent the basis for the ROC curves for the two markers implied by the model. Notice that the coefficient of D is the same for both ROC curves. In fact, the two equations differ only in their intercept. Therefore, the implied binormal parameters, a and b, are the same.

You can conclude that adding the covariate as a main effect only does not result in two ROC curves being modeled separately.

Now consider the following:

$$T = \alpha_0 + \alpha_1 D + \alpha_2 W + \alpha_3 DW + e^{2\beta D}\varepsilon$$

which produces the following submodels for $W=0$ and $W=1$:

$$T = \begin{cases} \alpha_0 + \alpha_1 D + e^{2\beta D}\varepsilon, & W = 0 \\ (\alpha_0 + \alpha_2) + (\alpha_1 + \alpha_3)D + e^{2\beta D}\varepsilon, & W = 1 \end{cases}$$

These two submodels generate two different ROC curves since the coefficient of D differs by a factor of α_3.

Specifically for $W=0$

$$a = \frac{\alpha_1}{e^{\beta}\sigma_0^2}, \quad b = e^{\beta}$$

and for $W=1$

$$a = \frac{\alpha_1 + \alpha_3}{e^{\beta}\sigma_0^2}, \quad b = e^{\beta}$$

Note that if $\alpha_3=0$, the ROC curves for the two markers are identical.

In principle this model can accommodate data collected on two markers. You can create a binary variable for any marker and then use it in place of *W* to perform the comparison. Typically, there are two practical differences between maker comparison and covariate adjustment:

- Marker comparisons usually involve paired data (as explained in the previous chapter), while covariate adjustments involve unpaired data.

- Marker comparisons usually use a *heteroscedastic* model (different variances allowed within the two groups defined by the outcome), while covariate adjustments usually employ a *homoscedastic* model (equal variances assumed within the two groups defined by the outcome).

Ordinal Predictors

5.1 Introduction

Chapter 3 addressed the single continuous predictor, probably the most natural and common setting for ROC curves. Nevertheless, the idea of an ROC curve applies equally well to an ordinal predictor. This chapter investigates how techniques from the previous chapter can be extended to accommodate an ordinal predictor.

Ordinal predictors can arise in a variety of ways. Sometimes, they are simply the result of categorization of an underlying continuous predictor. While this may result in the loss of information, and hence predictive accuracy, gains in having a simple and easy-to-communicate predictor may offset these losses. Rain forecasts might offer an example. Sometimes they are reported as low-medium-high, and other times they are reported in increments of 10%. Regardless of the terminology used to report the categorized predictors, data analysis can proceed using the predictors as ordinal.

Ordinal predictions also can arise through expert opinions. Most subjective opinion is hard to quantify, but it is relatively easier to offer a few scenarios ordered according to their probability—in effect, reporting ordinal predictions. Examples can be found in practically any field, such as medicine (diagnostic radiologists reporting their assessment of the chance or severity of disease), business (consultants opining on the likelihood of consumer preferences), or finance (stock picks, reported as buy, stay, or sell).

From the perspective of ROC curves, there is a duality between continuous and ordinal predictors. Both the empirical curve and the binormal model can accommodate ordinal predictors, either directly or with some modification. Sections 5.3 and 5.4 provide details.

5.2 Credit Rating Example

Consider the following data set collected and published by Güttler (2005) on the performance of the scoring systems of two of the most prominent credit evaluation agencies in the world: Moody's and Standard and Poor's. Table 5.1 reports the data on Moody's ratings, which use 17 categories on an ordinal scale ranging from Aaa (most favorable rating) to C (least favorable rating). The Rating row lists these 17 ratings, the Default row contains the number of companies that defaulted on their loans during the follow-up period of the study, and the Total row presents the total number of companies receiving the particular credit rating during the same period. The goal of this example is to evaluate the accuracy of Moody's ratings as predictors of the likelihood of default. For each of the 17 ratings, from Aaa to C, the total number in the sample and the number defaulting out of that total are provided.

Table 5.1 Moody's Ratings

Rating	Aaa	Aa1	Aa2	A1	A2	A3	A3-	Baa1	Baa2
Default	0	0	0	0	0	1	0	1	0
Total	49	47	120	164	163	254	226	238	242
Rating	Baa3	Ba1	Ba2	Ba3	B1	B2	B3	C	
Default	1	1	1	2	8	15	13	55	
Total	212	105	108	175	238	215	139	164	

Note that Table 5.1 contains all the data. With ordinal predictors, you can display an entire data set with nearly 3,000 observations because the space requirement has more to do with the number of categories than with the actual sample size. This is in contrast with continuous data, where the original data are usually not reported due to space constraints.

5.3 Empirical ROC Curve for Ordinal Predictors

The empirical ROC curve for an ordinal predictor is built on the same principle as the empirical ROC curve for a continuous predictor. In fact, because the empirical ROC curve is based on the ranks of the data only, whether the predictor is ordinal or continuous has no bearing on the way the ROC curves are constructed. Thus, the methods from Chapter 3 can be used to construct and compare empirical ROC curves with ordinal predictors.

To briefly review, using each distinct observed value of the ordinal predictor as a possible threshold, you can compute the sensitivity and specificity of the resulting binary predictor. A scatter plot of sensitivity versus one minus specificity for each threshold constitutes the ROC points. Note that the following code is identical to the program used in Chapter 3 for continuous predictors:

```
proc logistic data=moody;
   model default=moody / outroc=rocdata;
run;

axis1 length=12cm order=0 to 1 by 0.2 label=(f=swissb h=2)
value=(font=swissb h=2) offset=(.5 .5)cm;
axis2 length=12cm order=0 to 1 by 0.2 label=(a=90 f=swissb h=2)
value=(font=swissb h=2) offset=(.5 .5)cm;

symbol1 v=none i=join c=black;

proc gplot data=rocdata;
   plot _sensit_*_1mspec_ / haxis=axis1 vaxis=axis2;
run;
quit;
```

Figure 5.1 ROC Points for Moody's Rating Data

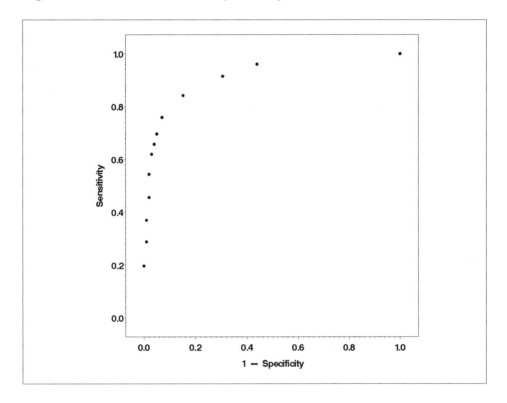

Figure 5.1 contains the ROC points for Moody's ratings obtained using PROC LOGISTIC. Note that, for this work, the variable `moody` has to be numeric. Also note the use of the OFFSET= option of the AXIS statement to print some white space on the two edges of the two axes for improved display. Also note that none of the companies receiving one of the first five ratings (AAA through A2) defaulted, so all five ratings are represented by the same point in the upper right corner (100% sensitivity and 100% specificity).

The term *ROC points* is used intentionally. Figure 5.1 does not have an ROC curve, only points each of which is a feasible operating point for Moody's ratings. It is possible to connect these points, just as we did for a continuous predictor, ending up with an ROC curve. You could argue that because unobserved intermediate thresholds are not possible by definition of the ordinal predictor, the resulting curve is not meaningful. Imagine that you connected the dots in Figure 5.1. There is no Moody's score between B2 and C, so the points on the line segment connecting the two leftmost points in the figure represent the operating characteristic (i.e., sensitivity and specificity) of thresholds that do not exist. The counterargument holds that the levels of most ordinal scales are arbitrarily defined and can be changed. If the ordinal predictions are obtained by categorizing a continuous predictor, then it is possible to explicitly revisit the thresholds. In the case of subjective opinions, experts can be asked to refine their predictions by using a scale with more levels. In our example, Moody's might decide, after examining Figure 5.1, that the gap between Aa1 and Aa2 (the second and third points from the right) is too wide. This could be resolved by introducing a new threshold between these two.

There is also a practical consequence of connecting the points: It makes it possible to unambiguously define the area under the curve for an ordinal predictor. Thus, I recommend using the ROC curve rather than ROC points for an ordinal predictor.

5.4 Area under the Empirical ROC Curve

The area under the empirical ROC curve is computed using the trapezoidal rule, the same way it was done for a continuous marker. The %ROC macro can be used for this purpose since the empirical ROC curve and the area under it are rank-based functions. The macro call is identical to the ones used in Chapter 3:

```
%roc(data=moody,var=moody,response=def,contrast=1);
```

The familiar %ROC output (see Output 5.1) indicates an AUC of 0.915 with a 95% confidence interval ranging from 0.891 to 0.940.

Output 5.1

```
The ROC Macro


      ROC Curve Areas and 95%
         Confidence Intervals
     ROC Area Std Error Confidence Limits

moody  0.9153   0.0124    0.8909     0.9397
```

Similar to using the %ROC macro, you can use the %BOOT1AUC macro to compute bootstrap-based estimates and standard errors for the area under the curve. Output 5.2 shows the results from the bootstrap, which are similar to those produced by the %ROC macro.

Output 5.2

```
Bootstrap Analysis of the AUC for moody

     AUC     StdErr
    _____

0.913584  0.012242



95% Confidence Interval

  LowerLimit     UpperLimit
  _____

   0.888271        0.935495
```

5.5 Latent Variable Model

As is the case with continuous predictors, the empirical ROC curve makes minimal assumptions but does not extend easily for covariate adjustments. In Chapter 3, we developed the binormal model and demonstrated its use in covariate adjustments in Chapter 4.

It is possible to develop a binormal model for ordinal predictors, too. Imagine that you start with a continuous predictor but instead decide to categorize it and report only an ordinal predictor, such as none, low, medium, and high (as in some of the rain forecasts). The consumers of the forecast only observe the ordinal value. Even though these continuous predictions, called the *latent variable*, remain unobserved, the sensitivity and specificity at each of the ordinal values represent points on the ROC curve of the latent variable. The fundamental idea is given in Figure 5.2.

Figure 5.2 The Three ROC Points for the Ordinal Predictor

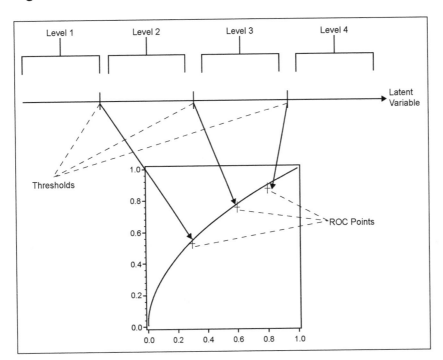

The horizontal line at the top portion of the figure represents the continuum of the latent variable. The three hash marks on the latent variable axis denote the thresholds that correspond to the observed ordinal predictor. The solid line at the lower part is the ROC curve for the latent variable and the hash marks on the curve are the ROC points. The important point is that the ROC points for the ordinal variable lie on the ROC curve for the latent variable.

Now examine Figure 5.3, which contains only the ROC points. Without knowing or supposing certain characteristics of the latent variable, you could not estimate the ROC curve of the latent variable by using only the ROC points. Connecting the points is unlikely to represent the true ROC curve of the latent variable because the true ROC curve is more likely to be smooth. You need to use information about the latent variable to interpolate between the ROC points. Of course, the latent variable is unobservable, so you can only make assumptions.

Figure 5.3 ROC Points for the Same Ordinal Predictor in Figure 5.2

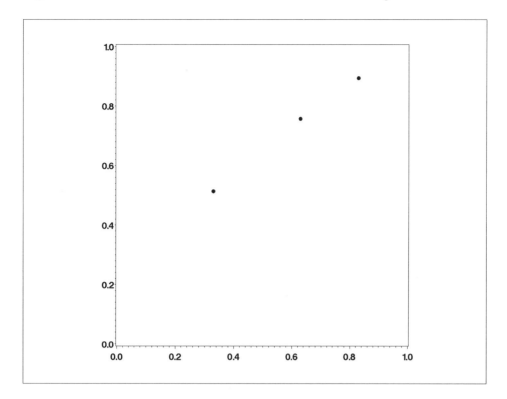

The most common assumption is that the latent variable is normally distributed. This gives rise to the so-called *latent binormal model*. For now, we will distinguish the latent binormal model from the one in Chapter 3. The latent binormal model assumes that the latent variable follows a normal distribution with mean μ_1 and variance σ_1^2 when $D=1$, and a normal distribution with mean μ_0 and variance σ_0^2 when $D=0$. It is not possible to simultaneously estimate μ_0, σ_0^2, μ_1, and σ_1^2 using the observed frequencies of the ordinal marker (such as those in Table 5.1). Multiple (in fact, infinite) combinations of these four parameters can give rise to the same observed data, and it is not possible to distinguish between them. However, if we anchor one of the normal distributions by fixing μ_0 and σ_0^2, then we can estimate μ_1 and σ_1^2. For simplicity, suppose we choose $\mu_0=0$ $\sigma_0^2=1$ and we get the estimates of μ_1 and σ_1^2, say $\hat{\mu}$ and $\hat{\sigma}^2$. Had we set $\mu_0=m$ $\sigma_0^2=s$, then the resulting estimates of μ_1 and σ_1^2 would have been $\hat{\mu}+m$ and $s\hat{\sigma}^2$. This can be proven because the normal distribution is a member of the so-called *location-scale families*. The implication is that, without loss of generality, we can always use the standard normal as the anchor and consider the resulting estimates as the difference of the two means and the ratio of the two variances.

These considerations led to the development of the following special ordinal regression model for ROC curves, first considered by Tosteson and Begg (1988). In this model, D is an indicator variable for the gold standard, that is $D=0$ or $D=1$. The latent variable X has a normal distribution with mean αD and variance $e^{2\beta D}$. This amounts to assuming a standard normal for $D=0$ and possibly a different normal density for $D=1$. Thus, this representation of X is consistent with the latent binormal model described previously.

Suppose that R is the ordinal predictor with k levels, created from X by applying the thresholds $\theta_1,..., \theta_\kappa$. Then the probability of R being equal to k is given by the following:

$$P(R = k) = P(\theta_{k-1} < X \le \theta_k)$$
$$= \Phi\left(\frac{\theta_k - \alpha D}{e^{\beta z}}\right) - \Phi\left(\frac{\theta_{k-1} - \alpha D}{e^{\beta z}}\right)$$

We saw in Chapter 3 that ROC curves are intimately connected to cumulative distributions. For this reason, using $\gamma_k(D) = P(R \le k)$ instead of $P(R=k)$ yields the following probit regression model:

$$\gamma_k(z) = \Phi\left(\frac{\theta_k - \alpha D}{e^{\beta z}}\right)$$

Realize that $1 - \gamma_k(D)$ is the probability of the marker exceeding the threshold and, hence, predicting $D=1$. This represents the sensitivity when $D=1$ and one minus the specificity when $D=0$. Therefore, a plot of $1 - \gamma_k(0)$ versus $1 - \gamma_k(1)$ for all k produces the ROC curve of the latent variable. The resulting smooth curve, analogous to the case of the continuous predictor, is

$$y = \Phi\left(\alpha + e^{-\beta}\Phi^{-1}(x)\right)$$

where x is one minus the specificity and y is the sensitivity. Under this model, the area under the curve is given by

$$A = \Phi^{-1}\left(\frac{\alpha}{\sqrt{1+e^{-2\beta}}}\right)$$

Setting $a=\alpha$ and $b=e^{-\beta}$, you can see that the formula for the AUC is equivalent to the case of a single continuous predictor. This is not surprising because we have continuously observed in Sections 5.2 and 5.3 that there is a strong duality between continuous and ordinal predictors.

We will call this an *ordinal-probit regression model*. Other ordinal regression models can be obtained by assuming a different distribution for the latent variable, in which case the probit link in this derivation would be replaced by the cumulative distribution function of the chosen distribution. In practice, it is very rare that another distribution is used in the latent model setting.

Those familiar with the capabilities of the LOGISTIC, GENMOD, and PROBIT procedures might mistakenly conclude that these procedures can serve as the primary vehicles for analyses of ordinal predictors. In fact, these three procedures are not flexible enough to accommodate the parameter β. Use them only if you are willing to fix β in advance at 1 and estimate only α. These procedures were developed to estimate a class of models called *generalized linear models*, and the ordinal-probit regression model defined here belongs to this family if and only if $\beta=1$. Therefore, to fit the ordinal probit model as described previously in full generality, you need to use the capabilities of PROC NLMIXED.

We will use Moody's data as an example. Note that the first five categories (Aaa through A2) have no observed defaults and, thus, the same sensitivity and specificity. In the language of this chapter, $\gamma_k(0)$ and $\gamma_k(1)$ are the same for $k=1, \ldots, 5$. Thus, we cannot estimate any of the four thresholds ($\theta_1, \theta_2, \theta_3, \theta_4$) because any of the points on the presumed latent variables distribution between $-\infty$ and θ_5 will be consistent with our data. To enable the ordinal regression to deal with all model parameters, we need to consolidate some of the categories. The consolidation in Table 5.2 was chosen for this analysis.

Table 5.2 Consolidated Version of Moody's Rating Data

Rating	A	Baa	Ba	B	C
Default	1	2	4	36	55
Total	1023	692	388	592	164

In the consolidated version, A represents Aaa through A3-, Baa represents Baa1 through Baa3, Ba represents Ba1 through Ba3, and B represents B1 through B3. C is not combined with another category.

Before we proceed with fitting this model it is important to emphasize the fact that the latent variable is not observed, so it is only the relative positions of the thresholds that can be estimated. In other words, two sets of thresholds with different absolute values but identical relative positions (i.e., the distance between them) will have the same likelihood. Hence we need to arbitrarily fix one of the thresholds so that others can be estimated. The following PROC NLMIXED code assumes the first threshold is 0 and fits the foregoing ordinal-probit regression model to the consolidated version of credit rating data.

```
proc nlmixed data=moody gconv=0;
  parms alpha=1 theta1=1 theta2=2 theta3=3 beta=1;
  bounds theta1>0, theta2>0, theta3>0;
  eta1=alpha*default;
  eta2=exp(beta*default);
      if rating=1 then z = probnorm(-eta1/eta2);
  else if rating=2 then z = probnorm((theta1-eta1)/eta2)
                          probnorm(-eta1/eta2);
  else if rating=3 then z = probnorm((theta2+theta1-eta1)/eta2)
                          - probnorm((theta1-eta1)/eta2);
  else if rating=4 then z = probnorm((theta3+theta2+theta1-eta1)/eta2)
                          - probnorm((theta2+theta1-eta1)/eta2);
  else if rating=5 then z = 1 - probnorm((theta3+theta2+theta1-eta1)/eta2);
  if z>1e-6 then ll=log(z);
  else ll=-1e6;
  model rating ~ general(ll);
  estimate 'AUC' probnorm(alpha/sqrt(1+beta**2));
run;
```

The short introduction to PROC NLMIXED in the appendix might be helpful if you are not familiar with the syntax or the output. Output 5.3 shows the results.

Output 5.3

```
                            Parameter Estimates

                    Standard
Parameter  Estimate    Error   DF  t Value  Pr > |t|  Alpha   Lower    Upper

alpha       1.9898    0.1157  2859   17.21    <.0001   0.05   1.7631   2.2166
theta1      0.5192    0.01679 2859   30.92    <.0001   0.05   0.4863   0.5521
theta2      0.3533    0.01667 2859   21.19    <.0001   0.05   0.3206   0.3860
theta3      0.9907    0.03948 2859   25.10    <.0001   0.05   0.9133   1.0681
beta       -0.2368    0.1336  2859   -1.77    0.0764   0.05  -0.4987   0.02517


Parameter Estimates

Parameter   Gradient

alpha       9.587E-7
theta1      0.000012
theta2      -2.58E-6
theta3      -3.04E-6
beta        -8.23E-6


                            Additional Estimates

                Standard
Label  Estimate    Error   DF  t Value  Pr > |t|   Alpha    Lower    Upper

AUC     0.9736   0.009248 2859  105.27    <.0001    0.05    0.9555   0.9917
```

The iteration history and the gradient are not useful to us as long as the model fitting procedure converges. Estimates of the most relevant model parameters can be found in the Parameter Estimates output section, with $\hat{\alpha} = 1.99$ and $\hat{\beta} = -0.24$. As discussed in Chapter 3, the ESTIMATE statement is a particularly convenient feature of PROC NLMIXED; it can compute the estimate and its standard error for any one-to-one function of the model parameters. Using this feature, we estimate the AUC to be 97%, with a standard error of 0.9% and a 95% confidence interval of 95.6% to 99.2%. By all accounts, Moody's ratings are excellent predictors of default.

Although not directly useful, the three thresholds are also estimated by this model. Their estimated values (labeled as theta1, theta2, and theta3 in the Parameter Estimates section) are 0, 0.5192, 0.5192+0.3533=0.8725 and 0.8725+0.9907=1.8632. Remembering that the latent variable is standard normal for $D=0$ and normal with mean 1.99 ($\hat{\alpha}$) and variance 0.62 ($e^{2\hat{\beta}}$), we can exactly compute the expected proportions in each level of the ordinal predictor.

Alternatively, we can compute the estimate of the AUC by simply plugging in the estimates of α and β, but calculation of the standard error requires some analytical legwork using the Taylor approximation.

The resulting ROC points are given as black dots in Figure 5.4, overlaid with the smooth line from the binormal model. The fit is very good for high thresholds (low sensitivity), but there are some deviations from the empirical curve for the higher thresholds.

Figure 5.4 The ROC Curve Implied by the Latent Variable Binormal Model

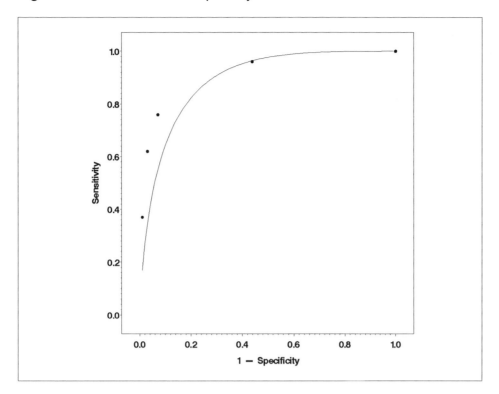

5.6 Comparing ROC Curves for Ordinal Markers

Comparing ordinal ROC curves can either be based on the empirical curve or the latent binormal model. The %ROC macro can be used to compare the areas under the two empirical ROC curves. Alternatively, the latent binormal model can be used to compare the parameters of the two latent binormal curves.

5.6.1 Metastatic Colorectal Cancer Example

Some of the tumors in the colon and rectum spread to the liver, in which case they are called *metastatic colorectal cancer*. If the spread is limited to the liver, then it can be removed surgically. But if there are other sites of metastases, then a liver operation is not indicated. For this reason, all metastatic colorectal cancer patients who are candidates for liver surgery are carefully screened for tumors outside the liver. The example in this section compares two different methods for extrahepatic imaging: computed tomography (CT) and positron emission tomography (PET). Each patient underwent both scans and each image was evaluated using a three-point scale by a radiologist: positive (presence of cancer), equivocal, and negative (absence of cancer).

5.6.2 Comparing the Areas under the Empirical ROC Curves

Just as empirical curves for an ordinal predictor can be constructed using the %ROC macro, areas under multiple empirical ROC curves can also be compared. The following call to the %ROC macro generates the desired comparison:

```
%roc(version,data=modif,var=pet ct,response=surg,contrast=1 -1);
```

Output 5.4 shows the results.

Output 5.4

```
                          The ROC Macro

                    ROC Curve Areas and 95%
                      Confidence Intervals
             ROC Area    Std Error   Confidence Limits

        pet  0.6744      0.2349      0.2141      1.1347
        ct   0.5526      0.2162      0.1289      0.9763


                    Contrast Coefficients
                                     pet      ct

             Row1                     1       -1

        Tests and 95% Confidence Intervals for Contrast Rows

   Estimate  Std Error Confidence Limits    Chi-square Pr > ChiSq

    0.6744    0.2349    0.2141    1.1347       8.2461  0.0041
    0.5526    0.2162    0.1289    0.9763       6.5346  0.0106


                    Contrast Test Results
                 Chi-Square   DF       Pr > ChiSq

                   11.6473    2          0.0030
```

The areas under the curve are 0.6744 for PET and 0.5526 for CT, with wide confidence intervals, in one case extending over 1. By definition, the AUC cannot exceed 1, so the upper limit of 1.1347 is a shortcoming of the asymptotic method that is used for confidence interval calculations. Generally, confidence intervals can be truncated from below at 0 and from above at 1. The difference between the empirical two areas under the ROC curves is 0.1218 and significant (p=0.003).

5.6.3 Comparing the Areas under the Latent Binormal ROC Curves

As an alternative to the empirical method, a latent binormal model might be assumed for both PET and CT. At this point, it is no surprise that the steps to compare the two latent binormal curves can be inspired by the binormal methods from Chapter 3. The key idea is to consider a regression model with the main effects of binary disease status (cancer/no cancer) and a binary indicator (PET/CT) as well their interaction. This leads to a model with a right-hand side identical to the models considered in Section 4.9. The coefficient of the interaction term is the key term again, representing the difference between the two ROC curves. The random term u represents the correlation between the two tests that is induced by the paired design.

The p-value for α_3 from the following code can be used to compare the two binormal curves. The variable Test is the binary indicator CT/PET.

```
proc nlmixed gconv=0;
   parms alpha1=1 alpha2=1 alpha3=1 theta2=3 beta=1 s=1;
   bounds theta2>0;
   eta1=alpha1*surg+alpha2*test+alpha3*surg*test+u;
   eta2=exp(beta*surg);
      if result=1 then z = probnorm(-eta1/eta2);
   else if result=3 then z = probnorm((theta2-eta1)/eta2) -
                             probnorm((theta2-eta1)/eta2);
   else if result=5 then z = 1 - probnorm((theta2-eta1)/eta2);
   if z>1e-6 then ll=log(z);
      else ll=-1e6;
   model result ~ general(ll);
   random u ~ normal(0,s) subject=patient;
run;
```

Covariate adjustments for ordinal markers can be done using the same representation in Chapter 4 by using an interaction term. There is no covariate of interest in the metastatic colorectal cancer example, but consider, hypothetically, that the age of the patient might have a bearing on the accuracy of CT scans. Then age-adjusted ROC curves for CT scans can be obtained using the following call to PROC NLMIXED:

```
proc nlmixed gconv=0;
   parms alpha1=1 alpha2=1 alpha3=1 theta2=3 beta=1 s=1;
   bounds theta2>0;
   eta1=alpha1*surg+alpha2*age+alpha3*surg*age;
   eta2=exp(beta*surg);
      if result=1 then z = probnorm(-eta1/eta2);
   else if result=3 then z = probnorm((theta2-eta1)/eta2) -
                             probnorm((theta2-eta1)/eta2);
   else if result=5 then z = 1 - probnorm((theta2-eta1)/eta2);
   if z>1e-6 then ll=log(z);
      else ll=-1e6;
   model result ~ general(ll);
run;
```

Note the absence of a random effect from u in this model. As we reviewed in Chapter 4, a notable difference between marker comparisons and covariate adjustment is that marker comparisons usually involve paired data (each patient measured by each marker) while covariate adjustment does not have this feature.

Lehmann Family of ROC Curves

6.1 Introduction

The first five chapters covered traditional methods of analyzing ROC curves. This chapter presents a different approach to constructing ROC curves for a continuous predictor. This approach has three important advantages in that it

- corresponds to an ROC curve with a very simple functional form

- easily fits with built-in SAS procedures

- lends itself to an easy generalization with respect to comparing ROC curves and covariate adjustments.

6.2 Lehmann Family of Distributions

In Chapter 3, we derived a general representation of the ROC curve in terms of the complement of the cumulative distribution function of the marker values conditional on the value of the gold standard status. If $D=0,1$ denotes the gold standard and W is the marker, then we defined

$$F(w) = P(W > w \mid D = 1)$$
$$G(w) = P(W > w \mid D = 0)$$

Then the ROC curve for W is given by

$$y = F\left(G^{-1}(x)\right)$$

This result holds in general and there is no restriction on F and G in particular. Chapter 3 used this result to derive the functional form of the ROC curve for the binormal model by simply replacing F and G with the corresponding cumulative densities of the normal distribution.

An alternative to the binormal family is to assume that F and G are related to one another as follows:

$$G(w) = F(w)^{\theta}$$

The family of distributions defined by this equation is called the *Lehmann family*. This family was used by Lehmann in his seminal work (Lehmann, 1953) on the study of the power function of various statistical tests. For this reason, this family of distributions is sometimes referred to as the *Lehmann alternative*.

It is important to realize that this equation does not uniquely define the actual distributions of marker values. It only identifies the distributions in reference to one another. For this reason, the term *semi-parametric* is commonly used to describe this family. Exponential distributions belong to the Lehmann family. Weibull distributions with a common scale parameter also belong to this family. On the other hand, normal and lognormal distributions are not members of the Lehmann family.

The previous equations imply the following:

$$y = x^{\theta}$$

which is the functional form of the ROC curve corresponding to the Lehmann family. In other words, if the conditional distributions of the marker values are linked to one another via θ in the Lehmann family, then the ROC curve will have the simple power form in this equation. Therefore, if you can estimate θ from the data, the estimate for an ROC curve will be immediately available.

An important advantage of the Lehmann family is that estimation methods for θ are widely studied and implemented. To see this, first define the following equation:

$$h_F(w) = \frac{f(w)}{F(w)}$$

where f is the density function corresponding to F and h is often called the *hazard function*, which plays a central role in survival analysis. If $G = F^{\theta}$, then the following relationship between the corresponding hazard functions holds:

$$\frac{h_F(w)}{h_G(w)} = \theta$$

Statisticians who routinely analyze censored data will recognize this as the family of proportional hazards. The Lehmann family and the proportional hazards family are identical. Cox (1972, 1975), in two key papers, developed the idea of partial likelihood and showed how it can be used to estimate θ. His work led the way to Cox (or proportional hazards) regression models, which have become the primary engine for modeling censored data. The PHREG procedure in SAS implements this methodology and provides an easily accessible venue for parameter estimation for the Lehmann family.

Although familiarity with the proportional hazards models and PROC PHREG might help you understand this chapter, it also might lead to confusion. Because proportional hazards models are almost exclusively used with censored data, it is easy to mistakenly assume that this chapter describes how to obtain ROC curves for censored data. In fact, neither the proportional hazards family nor the method of partial likelihood requires that the data be censored. The data sets in this chapter are actually identical in structure to the ones in Chapters 3 and 4: continuous markers and a binary gold standard. The goal is to provide an alternative to the binormal model, which has a full complement of regression models available.

The principle behind the regression versions of the binormal model developed in Chapter 4 was to use the marker value as the dependent variable and the gold standard as the independent variable in a regression model. We will use the same idea here. In the context of proportional hazards regression, this means

$$h_F(w; D) = h_G(w) \exp\{\beta D\}$$

PROC PHREG produces an estimate for β and its standard error. We can use these to obtain an estimate of θ and its standard error. Remember that W is the marker and D is the binary outcome. Therefore, if $D=0$, then $h_F=h_G$ and if $D=1$, then $h_F=h_G\, e^\beta$. Rearranging this gives the following equation:

$$\frac{h_F(w)}{h_G(w)} = e^\beta$$

Comparing this with $\dfrac{h_F(w)}{h_G(w)} = \theta$ gives the following:

$$\hat{\theta} = e^{\hat{\beta}}$$

You can use Taylor's theorem to approximate the variance of $\hat{\theta}$ as follows:

$$V\left(\hat{\theta}\right) = \hat{\theta}^2 V\left(\hat{\beta}\right)$$

Both $\hat{\beta}$ and $V\left(\hat{\beta}\right)$ are available from the PROC PHREG output.

Once θ is estimated, the ROC curve is completely specified and summary measures such as the AUC can be computed. Using the functional form of the Lehmann family of ROC curves, you can compute an estimate of the AUC as

$$A\hat{U}C = \int x^{\hat{\theta}} dx = \frac{1}{1+\hat{\theta}}$$

and the variance of the AUC estimate is given by

$$V(A\hat{U}C) = \frac{V(\hat{\theta})}{(\hat{\theta}+1)^4}$$

All the required components are again available from PROC PHREG output. You can also form pointwise confidence bands for ROC curves using the standard error of $x^{\hat{\theta}}$:

$$V(y(x)) = \left[x^{\hat{\theta}} \log x\right]^2 V(\hat{\theta})$$

Using the Lehmann family for ROC curves is studied in detail by Gönen and Heller (2007). ROC curves given by $y = x^{\theta}$ have been studied in the literature; see, for example, Hanley (1988). Their derivation using a Lehmann family and implementation using Cox regression procedures have been made explicit by Gönen and Heller (2007).

6.3 Magnetic Resonance Example

Zajick et al. (2005) reported a study on using chemical shift magnetic resonance in differentiating normal, benign, and malignant vertebral marrow processes. The marker of interest was the percent difference between the in-phase and out-phase signal intensities. The article focused on establishing a range of values for signal intensity change in normal vertebral marrow. Here, we use their data for a different objective: evaluating the ability of signal intensity change in discriminating between normal and benign vertebral marrow processes.

A total of 569 normal vertebrae were evaluated on 75 patients, compared with 215 benign lesions in 92 patients. Figure 6.1 presents the side-by-side histograms of the signal intensity change for normal and benign vertebrae separately. This figure is generated by the HISTOGRAM statement of the UNIVARIATE procedure. The complete code is given in Section 3.1 in the context of the positron emission tomography example. It is clear from the histograms that the two distributions have some overlap, suggesting, perhaps, that the marker may not have the ability to discriminate between the two classes.

Figure 6.1 Histograms of the Change in Signal Intensity for Normal and Benign Vertebrae

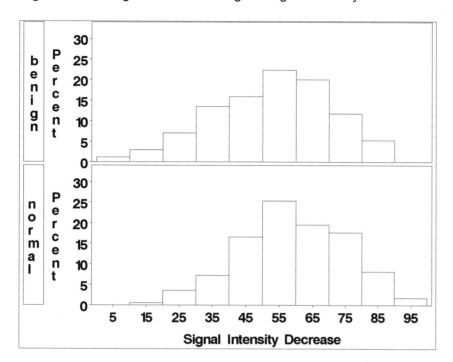

The empirical ROC points in Figure 6.2 verify this suspicion because the ROC curve is only slightly above the reference line. Figure 6.2 is generated by the OUTROC option in the MODEL statement of the LOGISTIC procedure using statements similar to the ones in Section 3.2.

Figure 6.2 ROC Points for the Change in Signal Intensity for Normal and Benign Vertebrae

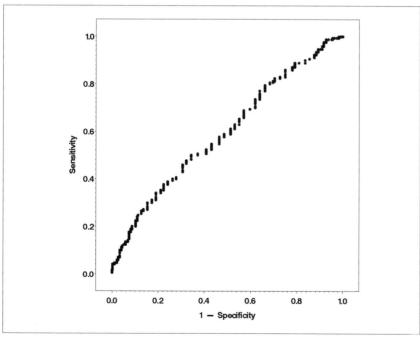

In the first analysis, ignore the fact that patients contribute multiple vertebrae to the analysis; this amounts to analyzing the 569 data points as if they originated from different subjects. To estimate θ, use the following call to PROC PHREG:

```
proc phreg data=tmp1.ph;
   model signal=grp / rl;
   if group='Normal' then grp=0;
   else if group='Benign' then grp=1;
run;
```

PROC PHREG is a SAS/STAT procedure that allows programming statements. These statements enable you to create variables that can be used in the MODEL statement. Using programming statements within the procedure helps avoid DATA step programming before the procedure is called. An important disadvantage is that the variables created are temporary; therefore, if the new variables will be needed in other places during the analysis, it is more efficient to use a DATA step.

In this example, programming statements are used to create a new variable called GRP. This variable is necessary because PROC PHREG works with numeric variables only. The RL option requests confidence intervals for the hazard ratio. Remember that θ, which is the only parameter in the Lehmann family, is the same as the hazard ratio.

The relevant part of Output 6.1 follows:

Output 6.1

```
                    Analysis of Maximum Likelihood Estimates

                   Parameter   Standard                            Hazard  95% Hazard Ratio
Parameter     DF    Estimate      Error  Chi-Square  Pr > ChiSq     Ratio  Confidence Limits

grp           1      0.35541    0.08815    16.2559       <.0001     1.427    1.200    1.696
```

Here, θ is estimated to be $1/1.427=0.701$. Whether θ is greater or less than 1 depends on the grouping order. If the following statements were used, the result would indicate that the hazard ratio is less than 1:

```
proc phreg data=tmp1.ph;
   model signal=grp / rl;
   if group='Normal' then grp=1;
   else if group='Benign' then grp=0;
run;
```

Remember that the hazard here is the chance of having a higher signal corresponding to grp=1.

The standard error of $\hat{\beta}$ is estimated to be 0.088 since the variance of $\hat{\theta}$ is given by $\hat{\theta}^2 \text{Var}(\hat{\theta})=0.062$. Using these, you can plot the Lehmann family ROC for this data set, along with 95% pointwise confidence bands (Figure 6.3).

Figure 6.3 The ROC Points (Dots), the Lehmann ROC Curve (Solid Line), and the 95% Confidence Intervals (Dashed Lines)

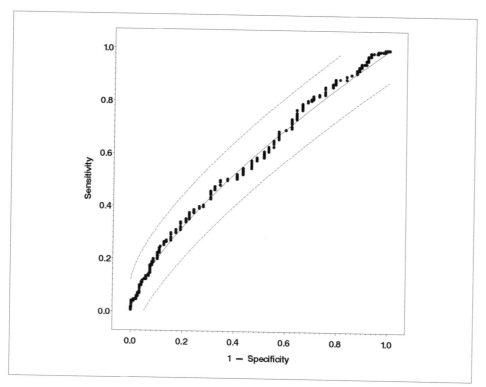

The area under the empirical curve is computed as 0.597, with a standard error of 0.025 (using the %ROC macro, details omitted), while the area under the Lehmann ROC curve is 0.588, with a standard error of 0.030. The latter pair of numbers is computed using the analytic relationships provided in Section 9.1. The closeness of the empirical and the Lehmann estimates is a sign that the Lehmann model is a good fit to the data.

6.4 Adjusting for Covariates

As we have seen earlier in Chapter 4, covariates may influence the accuracy of a marker and the ROC curve may need to be adjusted accordingly. It is easy to extend the regression models studied in Chapter 5 to the context of Cox regression, as follows:

$$h(V \mid D, U) = \tilde{h}(v) \exp\left\{\beta_1 D + \beta_2 U + \beta_3 DU\right\}$$

This provides regression-based covariate adjustments for the Lehmann ROC curves. This construction is similar in the sense that an interaction between the gold standard, D, and the covariate, U, is included in the model. To see the form of the covariate-adjusted ROC curve, consider the ratio of the hazard functions of observations, one with $D=1$ and one with $D=0$:

$$\frac{h(V \mid U, D=1)}{h(V \mid U, D=0)} = \exp\left\{\beta_1 + \beta_3 U\right\}$$

and the form of the adjusted ROC curve follows:

$$y = x^{\theta(u)}$$

where

$$\theta(U) = \beta_1 + \beta_3 U$$

Just as in Chapter 5, the ROC curve for the covariate is unaffected by β_2 but it is critical to include the main effect for age in the model; otherwise, the estimate of β_3, the key component for the ROC curve, might be biased.

Continuing with the magnetic resonance example, age may be an important factor in the accuracy of this type of magnetic resonance imaging, which focuses on the water-fat ratio in the tissue. Because patients naturally accumulate fat as they age, the marker may be less effective in the elderly, with the potential for mistaking fat due to age for fat due to a benign lesion.

The following PHREG call provides the required estimates:

```
proc phreg data=tmp1.ph;
   model signal=grp age grpage/rl;
   if group='normal' then grp=0;
   else if group='benign' then grp=1;
   grpage=grp*age;
run;
```

Again, note the use of programming statements, not just to create the group variable, but also the interaction. Output 6.2 contains the relevant numbers from the PHREG procedure. The three parameters are estimated to be β_1=1.29, β_2=-0.003, and β_3=-0.014. The negative sign of the coefficients for age indicate decreasing accuracy with increasing age. Output 6.2 shows the results.

Output 6.2

							95% Hazard	Ratio
		Parameter	Standard	Chi-		Hazard		
Variable	DF	Estimate	Error	Square	Pr > ChiSq	Ratio	Confidence	Limits
grp	1	1.28789	0.44629	8.3276	0.0039	3.625	1.512	8.694
Age	1	-0.00297	0.00320	0.8650	0.3523	0.997	0.991	1.003
grpage	1	-0.01432	0.00706	4.1123	0.0426	0.986	0.972	1.000

Analysis of Maximum Likelihood Estimates

Subsituting β_1 and β_3 in $\theta(u) = \beta_1 + \beta_3 U$ gives the following equation:

$$\theta(AGE) = 1.288 - 0.014 * AGE$$

which, when substituted in $y = x^{\theta(u)}$, gives the age-adjusted ROC curve:

$$y = x^{1.288 - 0.014 * AGE}$$

You can obtain a graphical representation by plotting the ROC curve for a few selected values of age (see Figure 6.4). Each ROC curve in Figure 6.4 is a plot of $y = x^{1.288 - 0.014 * AGE}$ for a different age value. The lowest curve represents 70-year-old patients and the highest curve represents 40-year-old patients. The curves in between represent varying levels of age in 5-year increments. As anticipated from the negative sign of β_3, the ROC curves get closer and closer to the diagonal line as age increases.

Figure 6.4 ROC Curves for Different Age Values

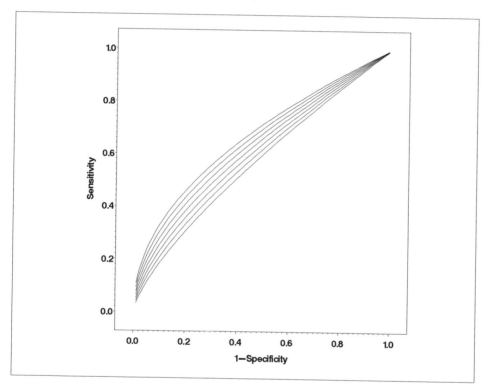

It is also possible to answer the question of whether a covariate adjustment is needed. Notice that when $\beta_3 = 0$, the adjusted and the unadjusted ROC curves coincide. Therefore, a test of $\beta_3 = 0$ is one way of answering whether you should report adjusted or unadjusted ROC curves. The *p*-value for this test is 0.0426 (see Output 6.2); thus, based on this analysis, an age-adjusted ROC curve is appropriate.

6.5 Using Estimating Equations to Handle Clustered Data

All the methods described so far require individual data points to be independent of each other. This requirement is usually satisfied when a single observation is made for each individual unit. Here, an individual unit can be a patient awaiting a diagnosis, a debtor applying for a loan, or a calendar day for which a weather forecast will be issued.

As was the case in the magnetic resonance example, sometimes more than one observation may be obtained from an individual unit. In this case, the research question centers on identifying abnormal vertebral marrow processes and classifying them as normal or benign. The radiologist performing the magnetic resonance imaging measures the signal at as many vertebrae as indicated based on the image. Some patients may contribute a single measurement while others will contribute several. When a patient contributes more than one observation to the analysis, the resulting data set is said to be *clustered*. The term is motivated by the observation that data coming from the same units behave similarly and tend to cluster together.

An intuitive way to think about clustered data is that the amount of information in two data points that are independent is generally more than the amount of information in two data points that are clustered. Clustering makes individual data points behave more like each other than they would otherwise. The net result is often an increase in standard errors (when compared with independence across data points), while point estimates are usually not substantially affected.

One can think of clustered data examples in other fields. A weather forecast for a given day is usually multi-dimensional: Temperature, precipitation, and pressure may all be simultaneously forecast. The forecasts within a given day will tend to be clustered because they are all related to various aspects of the same underlying meteorological events.

Clustered data do not always arise from multiple measurements on each observational unit. There could be other reasons to suspect that a subset of the observations behave like one another. For example, debtors in the same ZIP code may exhibit similar behaviors of default. Parts manufactured by the same factory may exhibit similar patterns of failure. All of these will have a bearing on the analysis of ROC curves.

Analysis of clustered data is a statistical topic in itself. An overview from a SAS user's perspective can be found in *SAS for Mixed Models, Second Edition* by Littell, et al. (2006) for continuous outcomes and *Categorical Data Analysis Using the SAS System, Second Edition* by Stokes, Davis, and Koch (2001).

One method of handling clustered data is based on generalized estimating equations. This is a way of estimating the model parameters using a method that is robust to departures from the assumption of the independence of data points. This method has been very popular in analyzing correlated binary data, and it has been extended to the case of censored data. It can be fit using the COVSANDWICH option in PROC PHREG. The option is named after the sandwich estimator of the covariance, which is one of the features of the estimators obtained by estimating equations.

If we modify the call to PROC PHREG as follows, the model is fit using estimating equations instead of the usual partial likelihood. The ID statement defines the variable that identifies the clustering variable. Records sharing the same ID value are clustered.

```
proc phreg data=tmp1.ph covsandwich(aggregate);
   model signal=group/rl;
   if group='Normal' then grp=0;
   else if group='Benign' then grp=1;
   id name;
run;
```

The format of the PHREG output is not altered with the use of the COVSANDWICH option. In particular we see that θ is estimated to be 1/1.427=0.701 (identical to Section 6.3), supporting the notion that there is very little effect, if any, of clustering on point estimates. Since $\hat{\theta}$ did not change, the estimate of the AUC remains the same at 0.588. The standard error of $\hat{\beta}$, on the other hand, is now 0.144, up from 0.088, an increase of 63.5%, as indicated by the StdErrRatio column, the only addition to PROC PHREG output in the presence of the COVS option. See Output 6.3. This results in an increase in the standard error of $\hat{\theta}$ to 0.101 and that of the AUC from 0.030 to 0.050.

Output 6.3

\multicolumn{7}{c}{Analysis of Maximum Likelihood Estimates}						
Parameter		DF	Parameter Estimate	Standard Error	StdErr Ratio	Chi-Square	Pr > ChiSq
group benign		1	0.35541	0.14413	1.635	6.0808	0.0137

\multicolumn{4}{c}{Analysis of Maximum Likelihood Estimates}			
Parameter	Hazard Ratio	95% Hazard Ratio Confidence Limits	Variable Label
group benign	1.427	1.076 1.892	group benign

Figure 6.5 adds to Figure 6.4 the confidence bands based on the estimating equations. The dotted lines both indicate 955 confidence intervals, one based on the standard error assuming independence (narrower bands) and one based on the estimating equations (wider bands). This figure is not intended to help you choose an estimation method. Whether you should use estimating equations depends on whether clustering can reasonably be expected based on the nature of the study.

Figure 6.5 The Influence of Clustering on the Variability of the ROC Curve

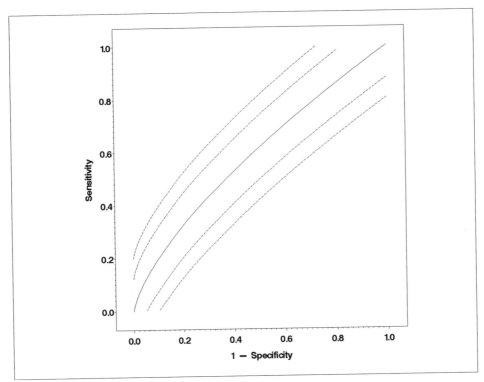

An attractive feature of estimating equations is that the method is based on regression models, so extensions to covariate adjustments are trivial both in principle and in implementation. For example, the following call to PROC PHREG fits an age-adjusted model, taking into account the fact that there are multiple vertebral marrow processes per patient:

```
proc phreg data=ph covs(aggregate);
   model signal=grp age grpage / rl;
   id name;
   if group='normal' then grp=0;
   else if group='benign' then grp=1;
   grpage=grp*age;
run;
```

The results, displayed in Output 6.4, again are presented similar to models fit using partial likelihood. The functional form of the age-adjusted ROC curve is identical to the one obtained in Section 6.4 and given by

$$y = x^{1.288-0.014*AGE}$$

The primary difference, as one might guess, is in standard errors. See Output 6.4.

Output 6.4

		Parameter	Standard	StdErr			Hazard	95% Hazard Ratio	
Variable	DF	Estimate	Error	Ratio	Chi-Square	Pr > ChiSq	Ratio	Confidence Limits	
grp	1	1.28789	0.89378	2.003	2.0763	0.1496	3.625	0.629	20.898
Age	1	-0.00297	0.00881	2.758	0.1137	0.7359	0.997	0.980	1.014
grpage	1	-0.01432	0.01337	1.894	1.1459	0.2844	0.986	0.960	1.012

Analysis of Maximum Likelihood Estimates

There is an increase of 89.4% in the standard error of β_3. This results in a p-value of 0.284, much larger when compared with the p-value from the partial likelihood analysis. We conclude that, after adjusting for clustering, there is no need for further adjustment on age.

6.6 Comparing Markers Using the Lehmann Family of ROC Curves

Chapter 5 explored the idea of using a regression model to compare two or more ROC curves. The principal idea was to include an indicator variable for the type of marker in a regression model. For the case of the binormal model, the appropriate regression model was a heteroscedastic normal regression. For the Lehmann family, the appropriate regression model is a proportional hazards regression, as we have already seen in Section 6.3.

Specifically, let U be an indicator variable that takes on values of 0 and 1 for the two different markers or predictors. Then, following the logic from Section 6.3, you can write the following equation:

$$\frac{h(V \mid U, D=1)}{h(V \mid U, D=0)} = \exp\{\beta_1 + \beta_3 U\}$$

This model is similar to those in Chapters 4 and 5 in the context of the binormal model. It is also identical in form to the one considered in Section 6.2, where U represents age, a continuous covariate. Therefore, the PHREG statements are also identical and are not shown here again.

If $\beta_3 = 0$ then the two ROC curves are identical. Therefore, a formal comparison can be done by testing $H_0 : \beta_3 = 0$. The Wald test readily available in PROC PHREG output can be used for this purpose. From an operational standpoint, comparing the ROC curves of two markers is no different than comparing the ROC curves for different age groups, so the steps described in Section 6.4 apply here, too.

There is a potentially important aspect in which marker comparisons may differ from covariate comparisons. As we discussed in Chapter 4, paired designs are the dominant form of data collection for most studies involving a comparison of multiple markers. The previous model, when estimated by partial likelihood, implicitly assumes that observations are independent. But Section 6.4 shows that an estimating equations approach to fitting the regression model allows you to obtain the proper standard error estimates when the data are correlated. Paired data are special cases of correlated data, and the marginal model approach outlined in the previous section can be used to fit the model in the presence of paired data.

6.7 Advantages and Disadvantages of the Lehmann Family of ROC Curves

With the advent of non-parametric methods for the ROC curve of a single predictor, you don't need to use the binormal model or the Lehmann family. The utility of these model-based approaches becomes evident only when you consider more complicated problems, such as covariate adjustments or correlated data.

As with any model-based statistical method, performance of the Lehmann family of ROC curves is a direct function of how well the data-generating process is approximated by the Lehmann assumption. If there is strong evidence that the data follow the Lehmann assumption, then the methods discussed in this chapter are appropriate and preferable. Under these conditions, commonly performed statistical tasks for ROC curves, such as covariate adjustment, comparison of several markers, and analysis using clustered data, can all be handled using PROC PHREG, a major advantage of the Lehmann family.

Despite the best efforts to verify or rule out the Lehmann assumption, the data might neither strongly deviate from nor strongly favor the Lehmann assumption. Although there is no substitute for careful and expert statistical analysis, you should consider the advantages and disadvantages for the Lehmann family and the competing methods. The chief advantages of the Lehmann family are its simple functional form, making it easy to interpret the results and gain insight, and the availability of the software. The primary disadvantage is its lack of flexibility: With a single parameter, the empirically evident shape of the ROC curve may not be accommodated.

ROC Curves with Censored Data

7.1 Introduction

In many applications, the binary outcome event is not immediately observable. For example, most credit scoring algorithms try to predict the probability of default by a certain time. If every subject in the data set is under observation at least until that time, then the outcome is truly binary and the methods we have seen so far are applicable. But it may not be desirable to wait until the outcome for all subjects is observed. It is possible to perform a time-to-event analysis, replacing the yes/no default with time elapsed until default. This analysis has the advantage of accommodating variable follow-up across subjects. Although it is not as powerful as waiting until all subjects reach the pre-specified time, it can usually be accomplished much quicker and the loss in efficiency is usually minimal.

7.2 Lung Cancer Example

Lung cancer is one of the most common and lethal cancers. Its prognosis is heavily influenced by two factors: tumor size and lymph node involvement. Using these factors, you can predict the likelihood of death and plan further treatment accordingly. These factors are best measured on the tumor specimen that is removed during surgery. However, there is considerable interest to accurately characterize the prognosis before surgery. Chapter 3 described the standardized uptake value (SUV) from a positron emission tomography (PET) scan. In this example we will examine the accuracy of SUV as a marker of lung cancer mortality. The goal of this analysis is to see if the SUV can predict survival in lung cancer patients following surgery. Since the SUV is available before surgery, it would have important practical consequences if it has reasonable predictive value.

The data set for this study has three variables: SUV, Survival, and Status. Survival is the time elapsed between surgery and death. For patients who are alive at the time of analysis, it represents the time between surgery and last follow-up. Status is a binary variable indicating whether the patient was dead or alive at last contact. If `status=1`, then the patient was dead and the Survival variable is the actual survival time. If `status=0`, then the patient was alive and Survival is the follow-up time. These patients are said to be *censored* because we do not observe their survival time.

This data structure is fairly typical of censored data, and it immediately reveals the intricate features of the required statistical analysis. The outcome (that is, the survival time) is represented by a combination of two variables (Survival and Status) in a specific way. Furthermore, for a subset of the patients, the actual outcome is not observed. Traditional statistical methods cannot be directly applied to the Survival variable for making inferences about survival time. For example, the median of the Survival variable will be an underestimate of the actual median because it treats the follow-up time of censored patients as if it was the actual survival time.

For this reason, analysis of censored data requires special methods. Because censored data are ubiquitous in clinical research, not to mention several other areas such as engineering reliability (time to equipment failure) and finance (time to default), these special methods have been widely studied. This chapter uses two of the more popular censored data methods. One is the Kaplan-Meier estimate of the survival time distribution and the other is the proportional hazards (Cox) regression. We have already made use of the Cox model in the previous chapter, although from a completely different perspective.

Strictly speaking, little prior knowledge about censored data is required to understand this chapter, but it would be difficult to grasp the details without some experience. Paul Allison's *Survival Analysis Using the SAS System: A Practical Guide* (1995) or Alan Cantor's *SAS Survival Analysis Techniques for Medical Research, Second Edition* (2003), both SAS Press books, can be helpful for this purpose.

7.3 ROC Curves with Censored Data

We have repeatedly emphasized that generating an ROC curve requires a binary outcome. If W is a predictor and D is the binary outcome, you could write

$$P(W > c \mid D = 1)$$
$$P(W \leq c \mid D = 0)$$

(7.1)

to denote the sensitivity and the specificity corresponding to a certain threshold, c. We have seen in Chapter 3 that these two probabilities can be written in terms of the conditional distributions of W and that they form the basis for the empirical ROC curve.

In the context of survival models, the outcome is the time elapsed until an event (such as death or default) takes place. This can be viewed as a binary outcome as a function of time. Equation (7.1) is now replaced by

$$P(W > c \mid D(t) = 1)$$
$$P(W \leq c \mid D(t) = 0)$$

(7.2)

which highlights the fact that sensitivity and specificity are functions of time in the context of censored data.

Using Equation (7.2), we can estimate the sensitivity and specificity for each c and plot these estimates to get the ROC curve at a specific time point, t. The estimates can be obtained using the following relations, which follow from the definition of conditional probability as well as application of the Bayes theorem:

$$P(W > c \mid D(t) = 1) = \frac{\{1 - S(t \mid W > c)\} P(W > c)}{1 - S(t)}$$

$$P(W \le c \mid D(t) = 0) = \frac{S(t \mid W \le c) P(W \le c)}{S(t)}$$

(7.3)

In Equation (7.3) and elsewhere in this chapter, $S(t)$ denotes the survival function—that is, $S(t) = P(T > t)$. It turns out that the three components on the right-hand side of Equation (7.3) can all be calculated using SAS. The following sections detail how Equation (7.3) can be computed using the data. For illustration, we will use $c = 9$ and $t = 36$, but the following steps can be repeated for other values of c and t as well.

7.3.1 Estimation of S(t)

As noted previously, $S(t) = P(T > t)$ is the survival function of the variable T (one minus the familiar cumulative distribution function), which is subject to censoring. Computing the cumulative distribution of a censored variable requires special methods. The most popular of these methods, alternatively known as the *Kaplan-Meier* or *product limit* method of estimation, is implemented in PROC LIFETEST in SAS/STAT software. PROC LIFETEST is the primary vehicle to compute the distributions required when using Equation (7.3). The following call returns the probability of survival at 3 years:

```
proc lifetest data=lung timelist=(36);
   time survival*status(0);
run;
```

The TIME statement has a special syntax, which combines information about the outcome from the two columns. The variable specified first (before the asterisk) is the survival time, and the variable specified second (after the asterisk) is the status. In parentheses after the status variable is a list of values that identify which values of the status variable indicate censoring. Finally, `timelist=(36)` prints out only the 3-year (36-month) estimates of the survival function. Otherwise, by default, survival probabilities for all time points observed in the data set are printed.

Output 7.1 shows the results of this invocation of PROC LIFETEST. The relevant portion is labeled Product-Limit Survival Estimates (product-limit is another name for the Kaplan-Meier estimate). The time point of interest is listed under the heading Timelist and the corresponding probability is labeled as Survival. Remember that $S(t) = P(T > t)$ and also note that $1 - S(t) = P(T < t)$ is given under the Failure heading. We see from the results in Output 7.1 that $S(t) = 0.7676$ and $1 - S(t) = 0.2324$.

Output 7.1

```
The SAS System
The LIFETEST Procedure

                     Product-Limit Survival Estimates

                                         Survival
                                         Standard    Number    Number
  Timelist   Survival    Survival  Failure  Error     Failed     Left

  36.0000    36.0000     0.7676   0.2324   0.0529      19        23


  Summary Statistics for Time Variable Survival

                  Quartile Estimates

               Point      95% Confidence Interval
               Estimate     [Lower      Upper)
  Percent

      75         .            .           .
      50         .         49.0000        .
      25      44.0000      24.0000        .


   Mean     Standard Error

  41.8379          1.4697

NOTE: The mean survival time and its standard error were underestimated
because the largest observation was censored and the estimation was restricted
to the largest event time.


  Summary of the Number of Censored and Uncensored Values

                                  Percent
    Total   Failed   Censored    Censored

     100      21        79        79.00
```

7.3.2 Estimation of $P(T|W)$

$P(T|W)$ is also, fundamentally, a time-to-event distribution and can again be estimated with PROC LIFETEST. The key point to remember is to exclude the appropriate patients based on the condition specified in W. Hence, when you compute $P(T|W>9)$, you need to include only those patients with an SUVgreater than 9, as shown in the following call to PROC LIFETEST:

```
proc lifetest data=lung(where=(suv>9)) timelist=(36);
   time survival*status(0);
run;
```

Output 7.2, which has the same format as Output 7.1 (with only the relevant portions shown), informs us that $1-S(T|W>9)=0.3665$. The equation $S(T|W\leq9)$, obtained similarly (though not shown here), equals 0.8960.

Output 7.2

```
The SAS System

The LIFETEST Procedure

                      Product-Limit Survival Estimates

                                        Survival
                                        Standard      Number
Number
Timelist    Survival    Survival    Failure    Error      Failed      Left

 36.0000    29.0000     0.6335     0.3665    0.0784       16          8
```

7.3.3 Estimation of *P(W)*

W is not subject to censoring, so, in principle, $P(W>c)$ is the proportion of observations exceeding c. $P(W)$ can be computed in many different ways in SAS, including using a DATA step or PROC SQL programming as well as using PROC UNIVARIATE and PROC FREQ. You can also use PROC LIFETEST because, in the absence of censoring, Kaplan-Meier methods produce the same results that would have been obtained from the standard methods. Because specifying a status variable is optional, PROC LIFETEST can be used for this purpose, as follows:

```
proc lifetest data=lung timelist=(9);
    time suv;
run;
```

This is a somewhat unusual call after the previous PROC LIFETEST calls. SUV is the variable for which a distribution is needed; hence, `timelist=9`. The lack of a status variable is due to the fact that W is not subject to censoring. The output from PROC LIFETEST (Output 7.3) indicates that $P(W>9)=0.50$. This is also the first example where the values under the headings Timelist and SUV differ. The former lists the value(s) requested by the TIMELIST= option of the PROC LIFETEST statement and the latter shows the nearest observed value for which the reported survival and failure probabilities hold. If the requested time point is observed, then the two columns will have identical numbers.

Output 7.3

```
The SAS System

The LIFETEST Procedure

                      Product-Limit Survival Estimates

                                        Survival
                                        Standard      Number
Number
Timelist      SUV       Survival    Failure    Error      Failed      Left

 9.00000    8.9300      0.5000     0.5000    0.0500       50          50
```

7.3.4 Putting It Together

Sections 7.3.1 through 7.3.3 demonstrate how the components of Equation (7.3) can be computed from the data using the SAS procedure PROC LIFETEST. Using the results of these sections, you can compute the sensitivity and specificity for an SUV of 9 at 3 years as follows:

$$P(W > c \mid D = 1) = \frac{\{1 - S(t \mid W > c)\}\, P(W > c)}{1 - S(t)} = \frac{(1 - 0.6335) * 0.5}{1 - 0.7676} = 0.7885$$

$$P(W \le c \mid D = 0) = \frac{S(t \mid W \le c) P(W \le c)}{S(t)} = \frac{0.8960 * 0.5}{0.7676} = 0.5836$$

By varying c, you can obtain the sensitivity and one minus specificity for each c. A plot of these pairs constitutes the ROC curve.

There is a potential problem here, however. In the examples presented in Chapter 3, the ROC curve was guaranteed to be monotone-increasing. There is no such guarantee for censored data because Kaplan-Meier estimates are not smooth functions of time (they have several *jumps*). This implies that, as one minus specificity increases, sensitivity might occasionally decrease, violating a central premise of the ROC curve. Lack of monotonicity may be obvious in small samples, but in most data sets with large samples and/or events, it is hardly noticeable.

If the estimate of $S(t)$ had no jumps and flat regions—that is, if it were monotone itself—the ROC curve would also have been monotone. Realizing this, Heagerty et al. (2000) suggest a different estimator for $S(t)$, a weighted Kaplan-Meier estimator.

The macro %TDROC generates a time-dependent ROC curve. The required inputs to the macro are DSN (the data set name), Marker, TimeVar, Status, and TimePT. The TimePT variable specifies the time at which predictions are to be made. The Status variable must satisfy the requirements of PROC LIFETEST: It must be numeric and it must be followed, in parentheses, by the list of values that indicate censoring. The macro also has an option (smooth=1) that implements a smooth estimator of $S(t)$. By default, smooth=0. Another optional input is PLOT (by default, 1), which controls whether the curve is plotted. When plot=0, only data sets with pairs of sensitivity and specificity are made available.

Figure 7.1 is an ROC curve for SUV as a marker of prognosis at 3 years using the lung cancer data generated by the following call to the %TDROC macro:

```
%TDROC(DSN=lung, MARKER=suv, TIMEVAR=survival, STATUS=status(0),
     TIMEPT=36);
```

Figure 7.1 The ROC Curve at 3 Years for the Predictive Power of SUV in Lung Cancer

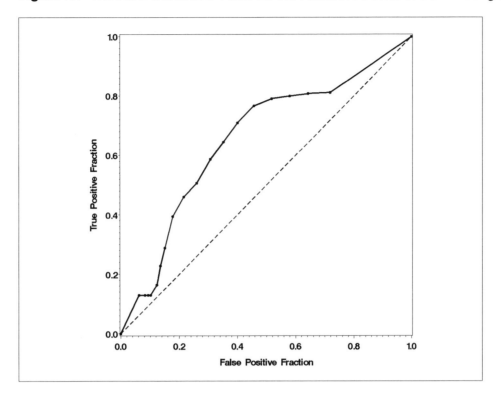

The AUC for the ROC curve in Figure 7.1 is 0.657 and suggests a moderate level of accuracy. The curve itself is never too far from the diagonal line, supporting the same conclusion. Of course, the ROC curve should be evaluated in the context of competing methods of prediction. In survival analysis, making accurate predictions is much more difficult because the event being predicted, in some cases, is several years away, during which many other things can happen to the patient. From this perspective, an AUC of 0.657 represents a respectable level of accuracy.

It is obvious that the choice of time point can influence the conclusions. In some cases, investigators have a clear target time point. Other studies lack such clarity and may pose a problem to the statistician in choosing the time point. In these cases, try out a few time points and present the results simultaneously, as in Figure 7.2.

Figure 7.2 The ROC Curves at 2, 3, and 4 Years for the Predictive Power of SUV in Lung Cancer

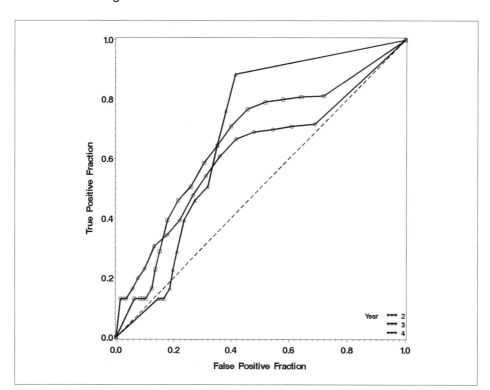

To generate a display like Figure 7.2, you need to understand how the output data sets are named by the %TDROC macro. The data set name in macro language is &&DSN_&TIMEPT, so for the lung cancer example, the data sets for the three ROC curves in Figure 7.2 are named LUNG_24, LUNG_36, and LUNG_48.

In general, the three curves have similar shapes, although sensitivity seems to increase with time at high levels of specificity and decrease with time at low levels of specificity. The AUCs of the three curves are 0.618, 0.657, and 0.729, suggesting that the SUV is better able to predict the status of patients at later years than earlier years. On the other hand, most of the difference between the curves is in the region where specificity is less than 0.5 and hence the difference in AUCs may be immaterial from a practical perspective.

7.4 Concordance Probability with Censored Data

Section 7.3 explained how to construct ROC curves with censored data. The principle idea is to dichotomize the time-to-event outcome at a given time point. As a result, the ROC curve is defined for a specific point in time. To get an idea about the overall predictive value of a marker, you need to perform an analysis like the one presented in Figure 7.2.

This section discusses an alternative method to assess the overall value of a marker in predicting a censored outcome. This alternative approach is based on the idea of concordance. In Chapter 3, we saw that, for a binary outcome, the area under the empirical ROC curve is equivalent to the concordance probability. As a reminder, the concordance probability is defined on a pair of subjects where one of the pair has the outcome and the other does not. The probability that the subject with the outcome has a greater marker value than the other subject is called the *concordance probability*.

The following way to express concordance probability is consistent with this definition and also makes it amenable to extend this definition to censored outcomes. Define

$$\psi\left(W_i, W_j\right) = \begin{cases} 1 & W_i > W_j \\ 0.5 & W_i = W_j \\ 0 & W_i < W_j \end{cases}$$

Hence Ψ indicates which member of the pair has the higher value, with ties indicated by 0.5. Suppose there are n patients with the outcome and m patients without the outcome. Then, there are a total of mn pairs and the concordance probability can be written as follows:

$$\frac{1}{mn} \sum_{i=1}^{n} \sum_{j=1}^{m} \psi\left(W_i, W_j\right)$$

The summation represents the number of pairs that have $W_i > W_j$ (with an accommodation for ties), so the entire expression is the fraction of patient pairs where the one with the higher marker value had the outcome.

You can use the idea of concordance in time-to-event settings. To see how the definition of concordance can be adopted for censored outcomes, let T_i and T_j be the event times in a given pair of patients with marker values W_1 and W_2. The concordance between a marker W and the censored outcome T is defined as

$$CP(W, T) = P(T_1 > T_2 \mid W_1 > W_2)$$

Because T is subject to censoring, this estimator cannot be used since the outcome is partially observed.

Harrell et al. (1982) suggest a modification of this estimator that can be used with censored data. This method is based on the realization that even in the presence of censoring, the outcome in some of the pairs can be ordered. For example, if the second subject is dead with a survival time that is shorter than the follow-up time of the first subject who is alive, we can say with certainty that $T_1 > T_2$. Table 7.1 lists all the possibilities and the corresponding value of $\Psi(W_1, W_2)$. If $\Psi(W_1, W_2) = 1$, it indicates that the prediction and the outcome are concordant. If $\Psi(W_1, W_2) = 0$, it indicates that the prediction and the outcome are discordant. Finally, $(W_1, W_2) = ?$ means that, due to censoring, it is not possible to determine whether the prediction and the outcome are concordant or discordant.

Table 7.1 All Possible Pairings and Concordance Status with Censored Data

Subject 1	Subject 2	T_1 and T_2	$\Psi(W_1, W_2)$
Event	Event	$T_1 > T_2$	1
Event	No Event	$T_1 > T_2$	?
No Event	Event	$T_1 > T_2$	1
No Event	No Event	$T_1 > T_2$	?
Event	Event	$T_1 < T_2$	0
Event	No Event	$T_1 < T_2$	0
No Event	Event	$T_1 < T_2$	?
No Event	No Event	$T_1 > T_2$	?

Now we can evaluate all the pairs to see whether they are concordant or not using the definitions in Table 1. The denominator is the number of informative pairs (that is the pairs for which we can make a decision), indicated in the following equation by k. The notation $(i,j)=1$ to $(i,j)=k$ means that the summation is over pairs of (i,j) and there are k such pairs. The resulting estimator of concordance probability is often called the c-index.

$$c = \frac{1}{k} \sum_{(i,j)=1}^{k} \psi\left(W_i, W_j\right)$$

The c-index has several advantages. It mimics the computation of the area under the ROC curve with binary data, it is intuitive, and it is conceptually simple to compute. Despite the conceptual simplicity, the computation requires forming all pairs and can take a long time. Computation of the c-index is not built into existing SAS procedures, but the %CINDEX macro, available from this book's Web site, can be used to compute it, as follows:

```
%CINDEX(DSN=LUNG,PRED=SUV,SURV=SURVIVAL,CENS=CENS);
```

Here, DSN is the name of the data set, Pred is the variable in the data set to be used as the predictor, Surv is the survival time variable, and Cens is the censoring indicator (1 for censored, 0 for event). Output 7.4 simply reports the c-index.

Output 7.4

```
cindex
_____

0.692708
```

In the case of the lung cancer example, the concordance probability is estimated to be 0.69, using the c-index as the estimator.

If the question of interest centers around multiple markers or a multivariable predictive model, then the input macro variable PRED of the %CINDEX macro should point to the variable containing the predictions of the multivariate model.

Despite all its advantages, the *c*-index has one big drawback. It is not a consistent estimator of the true concordance probability. A consistent estimate gets closer to the true value as the sample size increases. However, the *c*-index can remain far away from the true value even as the sample size increases. Therefore, the *c*-index can give "wrong" answers even in large samples. In my experience, if the censoring rate is low (which implies that the proportion of *undecided* pairs is low), the *c*-index works reasonably well. As the censoring rate increases, its performance becomes unpredictable.

The reason for the inconsistency of the *c*-index, at least intuitively, is that it uses the observed survival or follow-up times. Unless a parametric model is adopted, procedures using the observed values do not provide consistent estimates. For this reason, most of the non-parametric or semi-parametric methods of analyzing censored data, such as the Kaplan-Meier estimate or the proportional hazards regression, use the ranks of the data, not their actual values.

Before we finish this section, note an interesting property of the concordance probability:

$$CP(W,T) = 1 - CP(-W,T)$$

This equation tells us that if a marker is known to have high values when the outcome is unlikely, then *CP* can be less than 0.5. The more informative measure in this case is $1-CP$, which can be taken as a measure of the marker to correctly order event times.

7.5 Concordance Probability and the Cox Model

There is an alternative way of computing the concordance probability in a very important special case. Most cases involving regression with censored data, such as modeling the relationship between a continuous covariate and a censored outcome or between multiple covariates and a censored outcome, are handled within the framework of the Cox model (proportional hazards regression). It turns out that concordance probability can be computed analytically in the context of the Cox model. The following development is from Gönen and Heller (2005).

The starting point is the definition of concordance probability for a censored outcome, as used in Section 7.4:

$$CP(W,T) = P(T_1 > T_2 \mid W_1 > W_2)$$

where T is the survival time and W is the predictor. For the time being, we will work with a single predictor. Extension to the case of multiple predictors will follow. The Cox model for this case stipulates that

$$h(t|W) = h_0(t) \exp\{\beta W\}$$

where $h(t)$ is the hazard function for t, which is written as a product of two terms: one involving t only (h_0, the baseline hazard) and one involving W only (the linear part). Using the relationship between hazard and survival functions, you can write the equivalent expression:

$$S(t|W) = S_0(t)^{\exp\{\beta W\}}$$

More information on how to show the equivalence of these two expressions can be found in most books covering the Cox model, such as *Survival Analysis Using the SAS System: A Practical Guide* (Allison, 1995).

Now you can rewrite the concordance probability in terms of survival functions:

$$P(T_1 > T_2 \mid W_1 > W_2) = \int S_1(t \mid W_1) dS_2(t \mid W_2)$$

Understanding this equation requires some familiarity with advanced probability, but here it is used to show that concordance probability can be written using only $S(t)$. If you substitute the form of S_1 and S_2 from the Cox model, then the integral on the righthand side can be evaluated as follows:

$$P\left(T\left(\beta^T W_2\right) > T\left(\beta^T W_1\right)\right) = \int S_1(t \mid W_1) dS_2(t \mid W_2) = \left[1 + \exp\left\{\beta^T\left(W_2 - W_1\right)\right\}\right]^{-1}$$

Hence, to evaluate the discriminatory power of the predictor W in the context of a Cox model, you need an estimate of β and the values of W for all possible pairings of the data. Note that T enters the computation only through the estimation of β, which can be accomplished using partial likelihood. Therefore, censoring is naturally handled by the existing methods of fitting a Cox model. This is a particular strength of this approach to computing the concordance probability.

If there are multiple predictors, then both W and β will be vectors. The same expression can be written in vector notation as follows:

$$P\left(T\left(\beta^T W_2\right) > T\left(\beta^T W_1\right)\right) = \left[1 + \exp\left\{\beta^T\left(W_2 - W_1\right)\right\}\right]^{-1}$$

The expressions derived so far involve a particular pair of observations, denoted as 1 and 2. To use this in a set of data involving several observations, we need to compute this probability for each pair of observations. This yields the following formula for an estimate of the concordance probability (labeled CPE for concordance probability estimate):

$$CPE = \frac{2}{n(n-1)} \sum \sum_{i<j} \left\{ \frac{I\left(\hat{\beta}^T w_{ji} < 0\right)}{1 + \exp\left\{-\hat{\beta}^T w_{ji}\right\}} + \frac{I\left(\hat{\beta}^T w_{ji} \geq 0\right)}{1 + \exp\left\{\hat{\beta}^T w_{ji}\right\}} \right\}$$

where $w_{ji} = w_j - w_i$. The summand consists of two parts, representing the case whether higher or lower values of the predictor correlates with longer survival, which is reflected in the sign of the regression coefficient. If $\beta > 0$, then higher values of the covariate are associated with higher values of hazard. In this case, if, for a particular pair, $w_j > w_i$, then the probability of T_j being greater than T_i should be more than 0.5 and hence the second term of the summand should be applicable since $\beta^T w_{ji} > 0$ and $\exp\{\beta^T_{ji}\} > 0$. Thus, $(1 + \exp\{\beta^T_{ji}\})^{-1} > 0.5$. The other three cases depending on the sign of β and w_{ji} can be explained in a similar fashion.

As mentioned previously, the availability of Cox model software using partial likelihood is the primary operational advantage. It also constitutes one of the theoretical advantages. Partial likelihood estimates are *consistent*; that is, as the sample size grows, the results get nearer and nearer the underlying true but unknown value of the parameters. It is well-known that functions of consistent estimators are also consistent. Since β is the only parameter estimated to compute the CPE, it turns out that CPE is also consistent. This is an important reason why you should favor CPE over the *c*-index in the context of Cox models.

CPE can be estimated using the %CPE macro, which is available from this book's companion Web site at support.sas.com/gonen. The macro call is similar to that of %CINDEX:

```
%CPE(DSN=LUNG,COVARIATES=SUV SIZE,SURV=SURVIVAL, CENS=CENS);
```

Output 7.5 shows the results.

Output 7.5

Concordance Probability Estimate for the Cox Model	
CPE	StdErr
0.6524	0.0582

An important difference in the macro calls of CINDEX and CPE is that the predictors are listed with blanks in between as if they appeared on the right-hand side of the MODEL statement.

We see that the CPE for the lung data set is less than the *c*-index. This is usually the case, in my experience. The *c*-index seems to overestimate the true concordance probability, especially if the censoring proportion in high. Since the CPE is a consistent estimate and the *c*-index is not (as explained earlier), the CPE is a better measure in the context of using predictions from Cox regression models.

Using the ROC Curve to Evaluate Multivariable Prediction Models

8.1 Introduction

Our discussions so far have focused on a single predictor. Other variables were considered, such as when comparing several predictors or covariate-adjusted ROC curves, but not in the form of producing predictions from multivariable models and evaluating their accuracy. Chapters 8 and 9 deal with this problem.

It is useful to discuss the reasons for emphasizing single predictors. First, they provide the foundation for evaluating more complicated models. Second, single predictor applications are common in practice and thus deserve special focus. Diagnostic radiology is a good example where predictors are usually the products of patients undergoing scans. Although combining information from multiple sources of data (such as alternative diagnostic scans like magnetic resonance imaging and computed tomography) improves accuracy, the overwhelming financial and ethical concern is on minimizing the number of scans. This leads to concerns about picking the best single predictor out of a few candidates. Combining the information from multiple scan types is rarely of interest because the incremental improvement in accuracy does not usually justify the cost and burden for a patient to undergo additional diagnostic procedures.

Most prediction problems outside diagnostic radiology involve multiple variables. Usually, most outcomes of interest that you set out to predict are complex multidimensional entities that can be captured only through judicious use of several variables. This usually implies building a model, which requires choosing among several competing models. The goal of this process is finding the model that fits the data best, but it inevitably leads to over-fitting. *Over-fitting* refers to a model that describes the observed data much better than it anticipates future observations.

Measures of model performance (such as the ROC curve or the area under the ROC curve) that are computed from the data set used for model fitting are said to be obtained by *self-prediction* or *resubstitution*. These terms originate from the practice of using the model to predict the data that generated itself and resubstituting the model back into the data to obtain predictions. Another, somewhat light-hearted term for this method is *double dipping*, referring to the fact that the data set is used twice, once for fitting and once for predicting. The ROC curve and its summary measures tend to reflect the optimism in self-prediction because they indicate better accuracy than the actual model allows in practice.

Most statisticians recommend taking measures against over-fitting. This process is usually called *validation*. Here are two methods that are particularly relevant for ROC curves and that can be implemented in SAS with relative ease: split-sample validation and sample reuse validation. The sample reuse method can actually be implemented in two different ways: cross-validation and bootstrap. This chapter covers these methods.

Split-sample validation requires splitting the sample into two parts, so-called training and test sets. The model is fit on the training set and its performance is evaluated on the test set. This mimics the real-life situation where models are used on data sets that have not been part of the model development. Measures of performance estimated from the test set are much closer to the true values (less biased) than the ones estimated from the training set. Split-sample validation is simple to understand and implement. It also approximates real life well. Its main disadvantage is the inefficient use of data: with small to moderate sample sizes, neither the training nor the test sets are large enough to generate reproducible results.

Sample reuse methods attempt to increase efficiency by repeated use of observed data points in different ways. This repeated use results in estimators of performance with smaller variances. In general, however, because each data point is used more than once, there is an increase in bias. Nevertheless, most studies have favored sample reuse in terms of a composite criterion such as the mean squared error. In other words, the increase in bias is offset by the decrease in variance.

The most well-known sample reuse method is *cross-validation*. The data are divided into k segments (usually of equal size) and one part is set aside for testing while the remaining $k-1$ parts are used for training the model. Then this process is repeated for each segment and the resulting measures of accuracy are averaged over the k segments. When $k=n$, each data point constitutes a segment and the resulting process is called *leave-one-out* validation.

An alternative to cross-validation is to use bootstrap samples. We have seen in previous chapters how the bootstrap method can be used to obtain confidence intervals and *p*-values. When the goal is model validation, bootstrap samples obtained in the same way may be used to correct for over-fitting. Section 8.6 shows you how.

8.2 Liver Surgery Example

We will use a data set from the field of liver surgery throughout this chapter as an example. Surgery is the most promising treatment option for patients with liver cancer. The Liver data set has records from 554 surgeries that were performed to remove liver tumors. It is a subset of the data analyzed by Jarnagin et al. (2002). Variables include demographics, pre-operative patient characteristics such as existing co-morbid conditions, operative variables such as blood loss, and postoperative variables such as the incidence and severity of complications following the surgery.

Due to the nature of liver tumors, and aided by the fact that the liver has a unique ability to regenerate, most surgeries include removal of a substantial portion of the liver. This exposes patients to an elevated risk of complications. Predicting the likelihood of complications before surgery enables the treating team of physicians and nurses to increase post-operative monitoring of the patient as necessary. It is also helpful for counseling the patient and the patient's family.

We will use the following preoperative variables as potential predictors of post-operative complications: age, presence of any co-morbid conditions, extent of surgery (extensive vs. limited, where *extensive* is defined as an operation in which at least an entire lobe, or side, of the liver is resected), bilateral surgery (whether both lobes of the liver were involved in the resection or not), and number of segments resected (a segment is one of the eight anatomical divisions of the liver). Age and number of segments are considered as continuous variables. The typical number of resected segments is between 1 and 4, but occasionally 5 or 6 segments are taken out.

8.3 Resubstitution Estimate of the ROC Curve

The following code builds the predictions based on PROC LOGISTIC using a stepwise model selection method. Logistic regression is one of the many options available in SAS/STAT to build a predictive model for a binary outcome. Similarly, stepwise selection is one of several available model selection techniques. The goal here is not to claim logistic regression as the best way to build a predictive model, nor to promote stepwise as a model selection strategy. The emphasis is on model validation using several techniques.

```
proc logistic data=liver;
    model complications=age_at_op comorb lobeormore_code
    bilat_resec_code numsegs_resec /selection=stepwise;
run;
```

This invocation of PROC LOGISTIC requests a model to be selected using the stepwise method. Stepwise selection enters variables one by one and then removes them as needed according to their level of statistical significance. The following is an algorithmic summary of stepwise selection:

1. For each candidate variable, generate a *p*-value by adding the variable to the existing model from the previous step. In the first step, all variables in the model statement are candidates and the existing model is the intercept-only model.

2. Choose the candidate variable with the highest significance (lowest *p*-value) and update the existing model with the addition of the chosen variable. If no variables meet the criterion for entry ($p \leq 0.05$ by default but can be tailored using the SLE option of the MODEL statement), then the existing model is considered final.

3. Remove any variable that loses significance ($p > 0.05$ by default but can be tailored using the SLS option of the MODEL statement) from the existing model. If several variables have lost significance, remove each variable one by one starting with the variable with the highest *p*-value. After each removal, *p*-values are regenerated by refitting the model. Return the removed variables to the list of candidates.

4. At this point, all the variables in the existing model should be significant by the SLS criterion. All the other variables should be in the list of candidates. Go to Step 1 and repeat the process until none of the candidate variables are significant and none of the variables in the existing model can be removed.

Stepwise selection is a widely studied and somewhat controversial topic. It is used here to demonstrate the data-dependent nature of model selection and give a sense of how model selection procedures try to choose a model that provides the best fit to the data with little or no penalty for over-fitting. Although there are alternatives to or modifications of stepwise selection, generally, most forms of model selection lead to over-fit.

The relevant portion of the results appears in Output 8.1. The first table summarizes the stepwise selection. In this example, only two of the six variables were found to be significant: the number of segments resected and the age.

Output 8.1

```
                           Summary of Stepwise Selection

                 Effect              Number       Score        Wal
     Step   Entered      Removed    DF     In   Chi-Square  Chi-Square   Pr > ChiSq

       1   NUMSEGS_RESEC            1      1     52.6217                    <.0001
       2   AGE_AT_OP               1      2      7.9473                    0.0048


                 Analysis of Maximum Likelihood Estimates

                                      Standard        Wald
     Parameter        DF    Estimate     Error    Chi-Square    Pr > ChiSq

     Intercept         1      2.0959    0.3867      29.3791        <.0001
     AGE_AT_OP         1     -0.0166   0.00596       7.7810        0.0053
     NUMSEGS_RESEC     1     -0.3475    0.0518      45.0295        <.0001


                 Odds Ratio Estimates

                       Point        95% Wald
     Effect          Estimate   Confidence Limits

     AGE_AT_OP         0.984     0.972      0.995
     NUMSEGS_RESEC     0.706     0.638      0.782

     Association of Predicted Probabilities and Observed Responses

     Percent Concordant    69.4    Somers' D    0.392
     Percent Discordant    30.3    Gamma        0.393
     Percent Tied           0.3    Tau-a        0.196
     Pairs                76720    c            0.696
```

The familiar PROC LOGISTIC output indicates that this two-variable model has an AUC of 0.696, an estimate that we think is optimistic based on reasons described earlier. The ROC curve corresponding to this model is plotted in Figure 8.1. Of course, it is not just the area under the ROC curve that is overestimated; the estimated sensitivity for each value of specificity is optimistic, too.

Figure 8.1 Resubstitution Estimate of the ROC Curve for the Liver Resection Data

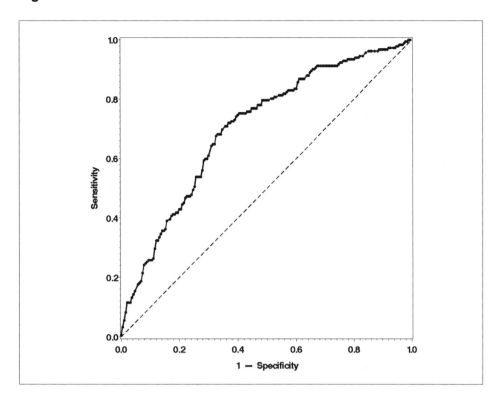

8.4 Split-Sample Estimates of the ROC Curve

Now, we'll see how to reduce the bias in resubstitution estimates using validation. The first step in split-sample validation is splitting the data set into two parts: training and test samples. They do not have to be of equal size. In fact, traditionally the training set is taken to be larger than the test set. A 2:1 split is commonly employed, where two-thirds of the observations are assigned to the training set and the remaining one-third to the test set. This split is usually done at random, although there are times when you might want to use a stratified split. A *stratified split* is one that ensures that proportionate amounts of one variable (or, at most, a few variables) appear in both training and test sets.

After the sample is split, the predictive model is built on the training set and estimates required for the ROC curve are obtained on the test set using the model optimized on the training set.

Splitting the sample can be done in many ways in SAS. The following DATA step is one example:

```
data training test;
   set liver;
   seed=12061996;
   unif=ranuni(seed);
   if unif>0.333 then output training;
   else output test;
run;
```

In this DATA step, a uniform random variable Unif is generated using the RANUNI function. Each observation is then assigned to either the training or test data set, depending on the corresponding realization of the Unif random variable. Essentially, the Unif variable represents a biased coin flip that determines whether the observation under consideration will be used for training the model or testing it.

The threshold for Unif is chosen as 0.333; therefore, approximately one-third of the records will have unif<0.333. Unif is a random variable, however, and each realization of this random variable produces a slightly different proportion of records where unif<0.333. For this reason, this method does not always produce an exact 2:1 match. If an exact 2:1 split is required, then you can sort the data on Unif first and then the desired number of records (in this example, exactly two-thirds of all records) for the training data can be taken from the beginning of the sorted sequence while the rest of the records are assigned to the test set. In most cases, an exact split is not necessary and the code given earlier will be sufficient.

After creating the training and test sets, the first task is to fit a model using the training data, as follows:

```
proc logistic data=training outmodel=model1;
   model complications=age_at_op comorb lobeormore_code
   bilat_resec_code numsegs_resec / selection=s;
run;
```

The primary difference in this call to PROC LOGISTIC is the OUTMODEL= option, which saves the details of the fitted model such as the parameter estimates and standard errors. The strength of the OUTMODEL= option is often more evident when it is combined with its sister option, INMODEL=, which accepts a model saved by the OUTMODEL= option as seen here:

```
proc logistic inmodel=model1;
   score data=test out=testscore;
run;
```

Note the absence of a MODEL statement. This is because a new model fit is not desired. Instead, all the calculations are based on the model specified by the INMODEL= option. The SCORE statement generates predictions on the test data set using the model fit specified in the INMODEL= option. The OUT= option of the SCORE statement generates a data set that contains predicted probabilities in addition to all the original variables in the data set specified in the DATA= option.

At this point, you can use the techniques developed in Chapter 3 to plot an ROC curve using the %PLOTROC macro or you can compute the area under the curve using the %ROC macro. The ROC curve and the area under it generated from the test set are considered unbiased estimates of the true ROC curve and the AUC. The key observation is that the predicted probabilities obtained by the SCORE statement serve as the marker values. For example, the following call to the %ROC macro produces the split-sample estimate of the AUC curve (see Output 8.2), which turns out to be 0.677, roughly 0.02 less than the resubstitution estimate of 0.696, illustrating the optimism in the resubstitution estimates:

```
%roc(data=testscore, var=p_1, response=complications, contrast=1,
      details=no, alpha=.05);
```

Output 8.2

```
The ROC Macro

        ROC Curve Areas and 95%
          Confidence Intervals
    ROC Area Std Error Confidence Limits

P_1  0.6767   0.0414    0.5957     0.7578
```

Note that P_1 is the variable name assigned to the predictions by the SCORE statement.

It is also possible to plot the split-sample ROC curve using the %PLOTROC macro introduced in Chapter 3. Since P_1 is given to several significant digits, a new variable P1 is created in the data set TESTSCORE to facilitate printing of thresholds. P1 is created as follows:

P1 = ROUND(P_1,0.01);

Then the following invocation of %PLOTROC is used:

```
%PLOTROC (dsn=testscore, makrer=p1, gold=complications, anno=4,
                      tlist=0.45 0.51 0.58);
```

Figure 8.2 Split-Sample Estimate of the ROC Curve Using the %PLOTROC Macro

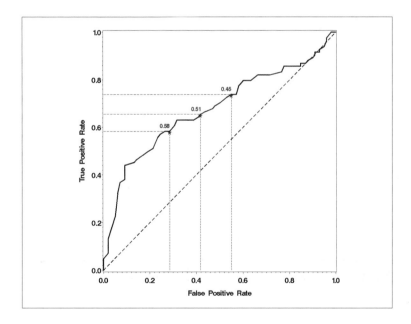

Note that thresholds in Figure 8.2 are described in terms of predicted probabilities because the marker plotted is the predicted probability from the logistic model (see MARKER= in the %PLOTROC macro call).

As an alternative to the %PLOTROC macro, you can use the OUTROC= option of the SCORE statement, which generates a data set that is identical in structure to the OUTROC= option in the MODEL statement. Specifically, it contains the variables _SENSIT_ and _1MSPEC_, which can be used in a PLOT statement of a call to PROC GPLOT.

```
proc logistic inmodel=model1;
    score data=test out=testscore outroc=rocvalid;
run;
```

Finally, it is instructive to overlay the resubstitution and split-sample estimates. This is accomplished in Figure 8.3, which shows the two ROC curves, one obtained by resubstitution and one by split-sample validation. The resubstitution curve is above the split-sample curve for the most part, but there are regions where the split-sample curve rises above the resubstitution curve. This seems contrary to the expectation that all the points on the ROC curve will be overestimated by resubstitution. This highlights the primary disadvantage of the split-sample approach; namely, it provides only one test set on which estimates of accuracy (such as sensitivity at a given specificity) can be computed. Although it provides unbiased estimates, they contain considerable variability and a situation like the one we have just observed is not uncommon. This is the primary reason for using the sample reuse methods discussed in the next two sections.

Figure 8.3 Comparison of the Resubstitution and Split-Sample Estimates of the ROC Curve for the Liver Resection Example

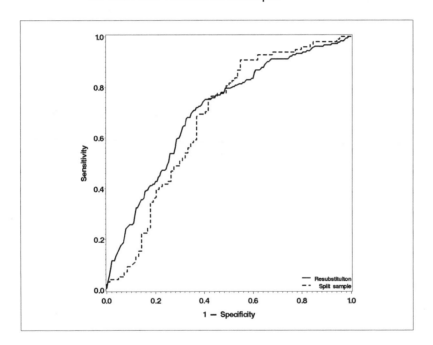

8.5 Cross-Validation Estimates of the ROC Curve

8.5.1 Leave-One-Out Estimation

PROC LOGISTIC offers a built-in option for cross-validation, which performs the leave-one-out method. As briefly explained in the introduction to this chapter, the *leave-one-out method* works by setting aside an observation, building a model on the rest of the data set, and then using the model to predict the left-out record. When this process is repeated for each observation, a prediction is obtained for every record in the data set using a model that was blind to the predicted observation. These predictions form the basis for a cross-validated ROC curve.

Leave-one-out cross-validation can be performed by PROC LOGISTIC using the following code. The PREDPROBS=X option in the OUTPUT statement specifies that leave-one-out cross-validation be used instead of resubstitution. This option applies to all the predicted probabilities generated by the OUTPUT statement. If the PREDPROBS= option is omitted, predicted probabilities are calculated using resubstitution.

```
proc logistic data=liver;
   model complications=age_at_op comorb lobeormore_code
      bilat_resec_code numsegs_resec / selection=s;
   output out=outx predprobs=x;
   ods output classification=classroc;
run;
```

Now the ROC curve can be plotted using the same methods recommended for split-sample validation, including one of the plotting macros. The area under the ROC curve and its standard error can be obtained through the %ROC macro. The following call to the %ROC macro produces the leave-one-out estimate of the AUC (Output 8.3), using the data set created by PROC LOGISTIC at the time the PREDPROBS=X option is specified. The variable name of the cross-validated predicted probability is XP_1:

```
%roc(data=outx, var=xp_1, response=complications, contrast=1);
```

Output 8.3

```
The ROC Macro


        ROC Curve Areas and 95%
           Confidence Intervals
     ROC Area Std Error Confidence Limits

XP_1  0.6875   0.0226    0.6431    0.7318
```

The estimate of the AUC, as obtained by leave-one-out cross-validation, is 0.6875 with confidence limits (0.6431–0.7318). Figure 8.4 contrasts the resubstitution estimate of the ROC curve and the leave-one-out estimate of the ROC curve. Although the two curves are very close, the leave-one-out estimate is consistently below the resubstitution estimate.

Figure 8.4 Comparison of the Resubstitution and Leave-One-Out Estimates of the ROC Curve for the Liver Resection Example

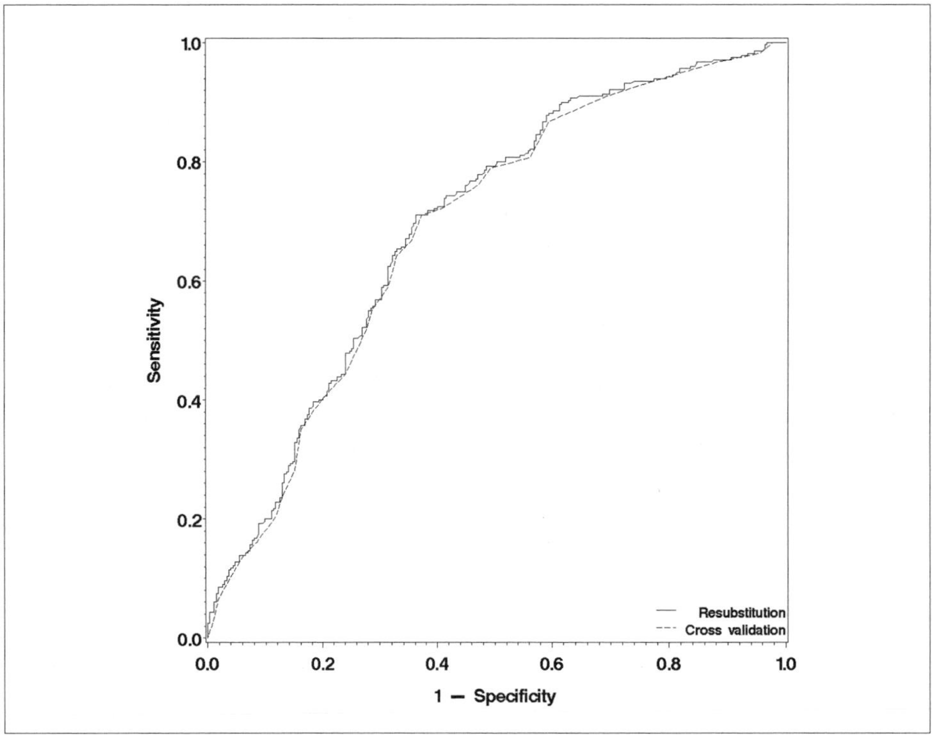

It may at first seem reasonable to use the OUTROC= option in this example to plot the ROC curve. However, the PREDPROBS=X option does not affect the data set created by the OUTROC= option in the MODEL statement. Even if you specify the OUTROC= and PREDPROBS=X options at the same time, the data set created by the OUTROC= option still reports the resubstitution estimates. Similarly, measures of association including c and Somers' D are also reported on a resubstitution basis. Therefore, the usual ways of obtaining ROC points and the area under the curve are not applicable for cross-validation, and these tasks have to be performed manually outside PROC LOGISTIC.

8.5.2 *K*-Fold Cross-Validation

While specifying the PREDPROBS=X option is a convenient way to obtain leave-one-out estimates, you should learn more about how cross-validation can be performed in SAS using the %XVAL macro. You can perform k-fold cross-validation using this macro. It is also helpful when making predictions with other procedures, such as the PROBIT, GENMOD, or GAM procedures, which do not offer standard cross-validation options.

The starting point for k-fold cross-validation involves breaking up the data in k randomly chosen segments, where k is a number supplied by the user. With k-fold cross-validation, the analyses are repeated k times. For each analysis, one of the k segments is left out as the test set and the remaining $k-1$ are used for training. In each analysis, the model is blinded to the test data. Table 8.1 explains the role of each segment cross-validation in the special case of $k=5$.

Table 8.1 The Role of Segments within Each Analysis in Five-Fold Cross-Validation

	Segment 1	Segment 2	Segment 3	Segment 4	Segment 5
Analysis 1	Training	Training	Training	Training	Test
Analysis 2	Training	Training	Training	Test	Training
Analysis 3	Training	Training	Test	Training	Training
Analysis 4	Training	Test	Training	Training	Training
Analysis 5	Test	Training	Training	Training	Training

There are a few advantages to this schema of validation:

- It is conceptually simple.

- Each segment (hence, each data point because segments are nonoverlapping) appears exactly once in the test set, so one and only one prediction is made for each data point.

- The choice of *k* allows for some flexibility in adjusting the size of the training and test sets.

The second point forms the basis of validation for our purposes. You can obtain an ROC curve directly from these predicted values since there is one prediction for each observation in the data set. This can be achieved by using any of the methods described in the previous section since, from the standpoint of obtaining ROC curves, there is no difference between a data set obtained by using the SCORE statement in PROC LOGISTIC and the test data combined from the *k* analyses from *k*-fold cross-validation.

The %XVAL macro, which is available from the book's companion Web site at support.sas.com/gonen, performs cross-validation. *K* is an input to the macro, along with the data set name, outcome, and covariates. The %XVAL macro produces no output but generates a data set called _XVAL_, which contains cross-validated predicted probabilities (P_0 and P_1=1−P_0) in addition to the original variables. This resulting data set can be used in calls to the %PLOTROC and %ROC macros to obtain cross-validated estimates of the ROC curve and the area under it. For example, these three macro calls perform a five-fold cross-validation for the Liver data:

```
%xval(dsn=liver,outcome=complications,covars=age_at_op comorb
         lobeormore_code bilat_resec_code numsegs_resec,k=5);
%plotroc (xval,P_1,COMPLICATIONS,anno=2);
%roc(data=xval,var=P_1,response=COMPLICATIONS,CONTRAST=1);
```

The ROC curve generated by the %PLOTROC macro appears in Figure 8.5 and the results from the %ROC macro appear in Output 8.4. The estimate of the five-fold validated AUC is 0.6883, with a standard error of 0.0226, virtually indistinguishable from the leave-one-out estimates.

Figure 8.5 Five-Fold Cross-Validated Estimates of the ROC Curve

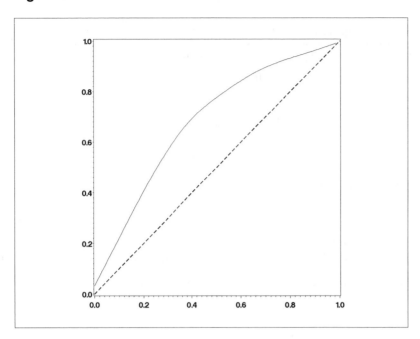

Output 8.4

```
The ROC Macro


        ROC Curve Areas and 95%
          Confidence Intervals
    ROC Area Std Error Confidence Limits

P_1  0.6883   0.0226     0.6440      0.7326
```

The need for cross-validation extends beyond linear logistic models fit by PROC LOGISTIC. Other SAS procedures that accept similar data but use different methods to fit models, such as PROC GENMOD, PROC GAM, and PROC PROBIT, can also benefit from cross-validation schema. It is difficult to write one general-purpose macro that will work with all the procedures because each one has different output syntax and Output Delivery System (ODS) table names. But it is relatively easy to tailor the %XVAL macro for different procedures. The initial DATA step in the macro adds a variable XV to the data set, which takes values from 1 to *K*, indicating which fold the observation belongs to. This portion is independent of the statistical prediction model or the SAS procedure used. You must modify the parts of the code that use PROC LOGISTIC. This modification requires procedure-specific expertise and may be accomplished by users who have experience with the particular procedure and with macro programming.

8.6 Bootstrap-Validated Estimates of the ROC Curve

Another sample reuse method of obtaining an ROC curve by validation is using bootstrap samples. This is not as intuitive as cross-validation, but it may, in certain cases, have a lower variance associated with estimation.

Bootstrap validation begins by forming *B* bootstrap samples. While for cross-validation, *K* was a small number (on the order of 5 to 10), for bootstrap validation, *B* will typically be larger, at least several hundred or perhaps 1,000. Each bootstrap sample is then used as a training sample. Then the original data set and the bootstrap sample are used as test samples. The difference between the measure of interest (say, the AUC) estimated from using the original data as the test sample and the measure of interest estimated from using the bootstrap sample as the test sample is a measure of *optimism*. This difference can be subtracted from the resubstitution estimate of the measure of interest to obtain the bootstrap-validated estimate. Estimates produced this way are sometimes called *optimism-corrected* or *optimism-adjusted* estimates.

The algorithm follows:

1. Compute the measure of interest, M, on the original data set, call it M_o.

2. Create B bootstrap samples. For $i=1, \ldots B$, do the following:

   a. Train a model on the bootstrap sample
   b. Compute M using the trained model on the original data; call it M_{oi}
   c. Compute M using the trained model on the bootstrap sample; call it M_{bi}
   d. Compute the difference: $d_i = M_{oi} - M_{bi}$

3. Subtract the mean(d_i) from M_o. This is the bootstrap-validated measure of interest.

The %BVAL macro, which is similar in syntax to the %XVAL macro, generates two data sets, BVAL1 and BVAL2, containing the predicted probabilities using the original data and the bootstrap sample, respectively. It also computes the optimistic AUC, optimism correction, and corrected AUC.

```
%bval(dsn=liver,outcome=complications,covars=age_at_op comorb
          lobeormore_code bilat_resec_code numsegs_resec,B=100);
```

Output 8.5 is obtained from a run of the %BVAL macro with 100 bootstrap samples.

Output 8.5

```
The SAS System

   Optimistic     Optimism
        AUC     Correction   CorrectedAUC

    0.695842     0.007179      0.688663
```

When the optimism correction of 0.007179 is subtracted from the resubstitution estimate of 0.695842, the bootstrap-validated estimate of the AUC is found to be 0.6887. Notice that this is virtually identical to the five-fold cross-validation estimate obtained in the previous section. In practice, with reasonably large data sets, leave-one-out, k-fold cross-validation, and bootstrap methods produce similar results, and the choice will be dictated by the traditions of the field of application as well as the accessibility of the software.

ROC Curves in SAS Enterprise Miner

9.1 Introduction

SAS Enterprise Miner is a collection of data mining tools. It was introduced in 1998 and used primarily by marketing analysts, database marketers, risk analysts, fraud investigators, and other professionals who try to find patterns in large volumes of data. SAS Enterprise Miner runs on a standard SAS platform, but it uses a graphical user interface (GUI) that is different from any other interface used by SAS. In addition, although the SAS engine on which it runs is accessible, SAS Enterprise Miner is designed with the premise that users will access it through its GUI. This contrasts with Base SAS and SAS/STAT, the focus of this book so far, which are accessed primarily through the program editor and by running code segments.

SAS Enterprise Miner offers a wide variety of methods for data mining, including clustering, dimensionality reduction, link analysis, and predictive modeling. This chapter focuses on predictive modeling. In Chapter 8, we studied ROC curves for multivariable predictive models, and logistic regression was reserved to demonstrate issues that arise in multivariable models and solutions using SAS/STAT. In many ways, this chapter is an extension of Chapter 8 because it uses the capabilities of SAS Enterprise Miner to examine classification trees and neural networks.

SAS Enterprise Miner has a built-in capability to generate ROC curves as part of its model assessment. In addition, you can create a SAS data set from the table that is created (behind the scenes) to generate the ROC curve. This data set can then be used with the techniques developed in earlier chapters for enhanced plots. Finally, you can score data sets within SAS Enterprise Miner using the predictive model(s) developed for this purpose. The results of scoring can also be saved as SAS data sets, enabling you to access the modeling methods of Chapters 3 through 6 as needed. In fact, it is even possible to generate SAS code to do the scoring.

This chapter is not intended to serve as a comprehensive guide to SAS Enterprise Miner. The intricacies of model building and selection, actions central to data mining, are presented only to the extent that their discussion helps you understand the resulting data curves. Various intermediate steps that we will use here, such as missing data imputation, may not be appropriate for some other data mining exercises. For a solid background on SAS Enterprise Miner, consult *Introduction to Data Mining Using SAS Enterprise Miner* by Patricia Cerrito (2006).

The audience for this chapter is SAS Enterprise Miner users, especially those who have not used ROC curves with the software. It is assumed here that you are already familiar with ROC curves at the Chapter 3 level, so a cursory reading of Chapter 3 may be useful at this point.

9.2 Home Equity Loan Example

The data for the example in this chapter are from the sample data library of SAS Enterprise Miner, Sampsio.Hmeq. The library Sampsio is internally defined when SAS Enterprise Miner is loaded and includes this data set. It contains nearly 6,000 records and 13 variables, one of which is the outcome variable, or target, in the language of data mining. The data set depicts the home equity loan records of a financial services company. Approximately 20% of the past home equity credit lines have resulted in default. In an attempt to reduce the default rate, or perhaps generate applicant-tailored interest rates, the company wants to build a model that predicts the risk of default on a home equity loan. The target variable is called BAD, with a value of 1 indicating a default and a 0 indicating a serviced loan. The 12 predictor variables, along with a brief definition, are as follows:

- CLAGE: age (months) of oldest credit line
- CLNO: number of credit lines
- DEBTINC: debt-to-income ratio
- DELINQ: number of delinquent credit lines
- DEROG: number of major derogatory reports
- JOB: occupation of the debtor
- LOAN: amount of the loan
- MORTDUE: amount due on existing mortgage
- NINQ: number of recent credit inquiries
- REASON: reason for loan (debt consolidation or home improvement)
- VALUE: value of the property
- YOJ: years at present job

The goal is to find a predictive model for the BAD variable using the 12 other variables in the data set. Although we will go through the steps required to build a predictive model, the focus is on constructing ROC curves using SAS Enterprise Miner's built-in functionality, as well as exploring ways to export the predicted values to a SAS data set so that the methods and macros discussed in earlier chapters can be used to perform custom analyses.

9.3 ROC Curves from SAS Enterprise Miner for a Single Model

SAS Enterprise Miner organizes its objects in a flow diagram, which contains data, models, and results. The components of the diagram are called *nodes* and they can be connected to other nodes in the flow. The flow can be saved as a project, which may contain other flows as well.

Most actions in SAS Enterprise Miner can be specified by three equivalent methods: using the Menu, using the Tool Bar, or using the Project Navigator (the left segment of the window when SAS Enterprise Miner first starts). The Tool Bar and the Project Navigator contain icons that can be dragged and dropped, so they are more convenient than using the Menu. You can also customize these icons.

Follow these steps to prepare the data for predictive modeling. First, define the sources of data. This is similar to using LIBNAME statements in a SAS DATA step and then using PROC COPY to bring the desired data sets to the work area. Then, choose a specific data set to work with from the data sources. Partitioning the data into training, validation, and test data sets is customary in data mining and can be accomplished using the Data Partition node. Training and validation data are discussed in Chapter 8, where the term *test data* is used interchangeably with validation data. The terminology in this chapter follows that of SAS Enterprise Miner, so validation and test data here refer to two different entities. The validation data set is used for the same purpose it is in Chapter 8, but the test data set here is primarily held out for a confirmation of the findings of the validation data set. This practice is less common in areas other than data mining.

The following steps more precisely describe how to perform these tasks:

1. Define data sources by right-clicking **Data Sources** in the Project Navigator, choosing **Create Data Source**, and proceeding through the menus. Make sure that the **Model Role** field for the variable BAD is Target and the **Type** is binary. All other variables should be listed as Input.

2. Create a new diagram by dragging and dropping from the Tool Bar or the Project Navigator. The Diagram icon depicts a series of interconnected boxes with a yellow star above them. Alternative ways to create a new diagram include selecting **File→New→Diagram** or holding down the CTRL and Shift keys and pressing D.

3. Double-click **Data Sources** in the upper left segment of the window. Drag and drop Home Equity into the diagram workspace (the right segment of the window, which should be blank at this point).

4. Add Data Partition to the diagram by dragging and dropping the icon into the diagram workspace or alternatively selecting **Action→Add Node→Sample→Data Partition**.

5. Connect the Home Equity and Data Partition icons with an arrow. You can draw an arrow when the cursor is shaped like a pencil.

Note that the diagram specifies the actions to be undertaken, but merely specifying them is not enough for execution. A flow, either as a whole or as a subset, needs to be run for the specified actions to be carried out. For example when the Data Partition node is run, training, validation, and test sets are created. To run a flow, right-click on a node and choose **Run**. All the nodes prior to and including the chosen node execute in sequence. It is possible to run each node as the diagram is specified, but it is more efficient to run the flow in big chunks.

Now we are ready to proceed with modeling. Since the target variable is binary, a reasonable starting point for modeling is logistic regression. To fit a logistic model in SAS Enterprise Miner, you need to add a node that specifies this and connect the regression node to the data sets created by the Data Partition node. The following steps describe this process:

1. Add the Regression node to the diagram by dragging the icon when the **Model** tab is highlighted.

2. Connect the Data Partition node to the Regression node with an arrow.

3. Add the Model Comparison node to the diagram by dragging and dropping it from the **Assess** tab and connecting it to the Regression node with an arrow.

At this point, the display should roughly look like Figure 9.1.

Figure 9.1 Using Logistic Regression to Predict Defaults on Home Equity Loans

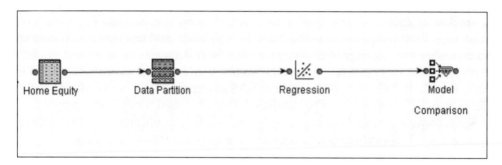

Now, right-click the Model Comparison node and choose **Run**. Once a node has run, the results remain available throughout the SAS Enterprise Miner session. The diagram and the results are saved by default and are available when the SAS Enterprise Miner session is restarted later.

When the Model Comparison node has run, right-click on the node and choose **Results**. A new window appears, with subwindows titled Fit Statistics, Output, and ROC Chart. In the ROC Chart subwindow, there are three ROC curves, one for each data set: training, validation, and test. Figure 9.2 focuses only on the ROC curve for the training set, displayed in the upper left corner. The 45-degree reference line, referred to as Baseline in the legend, is plotted by default. As you move the cursor along the curve, a pop-up text box provides details about the sensitivity and specificity at that point of the ROC curve. Figure 9.2 includes one example point where the sensitivity is 50.5% and the one minus specificity is 94.9%.

Figure 9.2 ROC Curve for the Logistic Regression Model Using Home Equity Loans Data

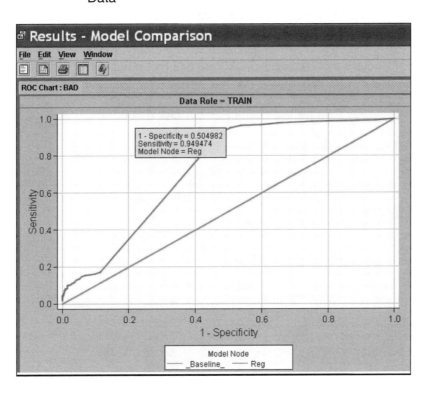

As Output 9.1 shows, the areas under the curve are readily reported in the Output window, labeled as ROC index, as 0.72385, 0.69319 and 0.67768, respectively for the training, validation, and test sets.

Output 9.1

```
Fit Statistics
Model selection based on _TMISC_

Train:       Valid:       Test:
Roc          Roc          Roc
Index        Index        Index
0.72385      0.69319      0.67768
```

9.4 ROC Curves from SAS Enterprise Miner for Competing Models

In addition to logistic regression, we will also use two other methods in the toolkit of SAS Enterprise Miner: decision trees and neural networks. They are not available in SAS/STAT, so a statistician who uses SAS may be unfamiliar with them. For this reason, here's a brief description of these methods.

Decision trees, also known as *recursive partitioning* in statistical literature, mostly consider binary splits of each variable in a hierarchical manner. For the first split, all the variables are considered one by one and the one that provides the best split (according to one of the several commonly used criteria, such as the misclassification rate or Gini index) is retained for the first step. Nodes that are not split any further are called *terminal nodes* or *leaves*. Once this variable is chosen, the same process is repeated again twice, one for each side of the binary split. SAS Enterprise Miner also considers multiple splits of one side of the node so that it is possible to have a tree where one side of the initial node is a leaf and the other side has split multiple times. Multinomial variables are grouped into two so that the resulting binary grouping gives the best split. Similarly, the continuous variables are also split into two by finding the best cutoff. The resulting decision tree for the home equity loan example appears in Figure 9.3.

Figure 9.3 Example Decision Tree for the Home Equity Loan Example

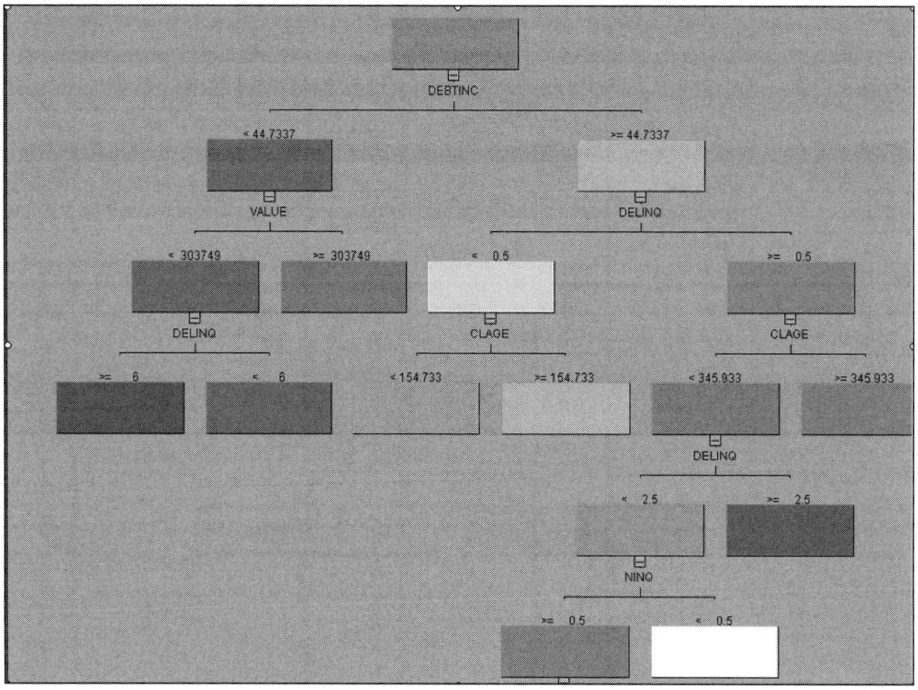

For example, the first variable used is DEBTINC, which is a continuous variable. It was split at 44.7337. For those with a debt-to-income ratio less than 44.7337, the next best split was VALUE, split at 303,749.

Decision trees are distinguished from logistic regression and neural networks in an important way: There is no statistical model. Because there is no model, there is no predictive equation. When a new observation is scored, it is "dropped" from the top of the tree and follows the route that is dictated, at each node, according to the value of the covariates until it reaches a terminal node (or leaf). This can be described as a rule, in the form of a series of IF-THEN-ELSE statements.

The binary split structure of the tree is an advantage because it is easy to interpret. It can also be a disadvantage, especially when some of the predictors are continuous. Categorization of continuous covariates may lead to inefficiency and adversely affect predictive performance. Another advantage of decision trees is that they handle missing data very effectively. This is accomplished by establishing surrogate splits at each node that can be used when the splitting variable for that node is missing.

Neural network is an umbrella term that describes methods that build models mimicking the function of the brain. Their precise mathematical description is complicated and unlikely to benefit an ROC-focused discussion. Think of neural networks as a specific yet flexible class of nonlinear regression models.

Before we run decision trees and neural networks on the same data set, there is one important issue to consider about missing data. Both logistic regression and neural networks require complete observations. If any portion of a record is missing, that record is completely ignored during model building. Decision trees are more forgiving. They consider surrogate splits in case the variable for the primary split is missing. For this reason, if we run logistic regression, neural networks, and decision trees on the same input data, the actual data that the predictive model rests on may not be identical from one method to the other. You could address this by imputing the value from a distribution derived from the normal density. An imputation node can be dragged and dropped from the **Modify** tab, as seen in Figure 9.4. The same figure also shows the addition of decision tree and neural network nodes, both dragged and dropped from the **Model** tab.

Figure 9.4 SAS Enterprise Miner Diagram Workspace for the Competing Models Example

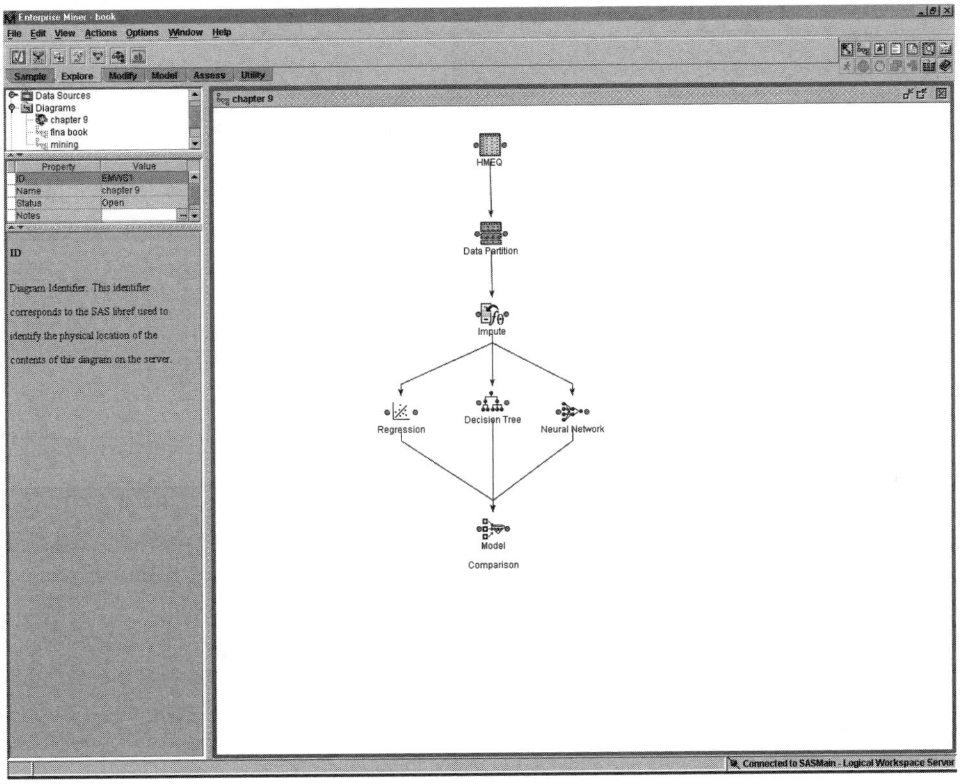

The flow in Figure 9.4 implies that the same data set will be analyzed three different ways and the results compared in the final node.

Three ROC curves, one for each method of analysis, are plotted to better assess the predictive power of the models (see Figure 9.5). Neural networks and logistic regression have very similar performance. The decision tree outperforms regression and neural networks. It has higher sensitivity than the other two for a given value of specificity for the most part, especially in the range where both sensitivity and specificity are above 0.5. The neural network method is slightly better than the regression method, although the difference between the two seems to dissipate in the validation and test sets. Output 9.2 reports the AUCs for each model and data set. The AUC for the decision tree is consistently above 85%, while the other two models' AUCs cluster around 80%.

Output 9.2

Model Node	Train: Roc Index	Valid: Roc Index	Test: Roc Index
Neural	0.84003	0.78812	0.80917
Reg	0.81356	0.78147	0.79139
Tree	0.86871	0.85724	0.85971

Figure 9.5 ROC Curves from SAS Enterprise Miner for Each of the Three Models and Data Sets

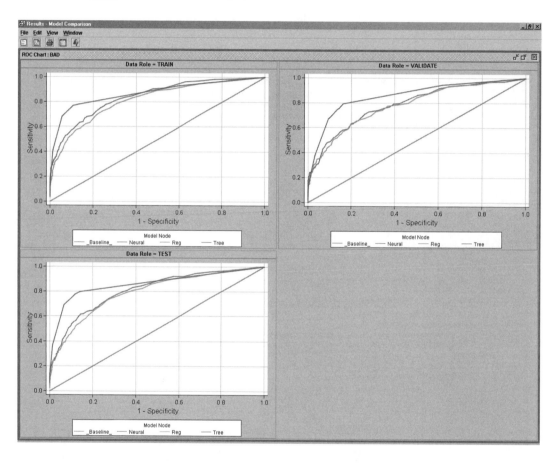

9.5 ROC Curves Using PROC GPLOT with Exported Data from SAS Enterprise Miner

While SAS Enterprise Miner provides a built-in capability for plotting the ROC curve, sometimes you might want to output the data set containing the ROC points so that you can use the %PLOTROC macro for custom graphics. This can be done by selecting **View→Table**, which displays the SAS data set behind the ROC curve. At this point, selecting **File→Save As** gives you the option to save the table as a permanent data set that can be accessed later using Base SAS and SAS/GRAPH.

To give you some ideas, Output 9.3 lists the first 23 observations in the data set produced by SAS Enterprise Miner using the flow in Figure 9.4. The key variables are SENSITIVITY and ONEMINUSSPECIFICITY. Additional useful variables are MODELNODE and DATAROLE. The MODEL variable contains four categories: _Baseline_ (denoting the model where no covariate information is used, which results in the 45-degree line), Neural (corresponding to the Neural Network node), Reg (corresponding to the Regression node), and Tree (corresponding to the Decision Tree node). DATAROLE simply indicates whether the corresponding observation is for training, validation, or testing only. Traditionally, ROC curves are plotted only for the validation data set, but you can compare the amount of overestimation with the availability of this variable in Output 9.3.

Output 9.3

Obs	Node	Model Node	Variable	Role	Event	Specificity	Sensitivity
1	Neural	_Baseline_	BAD	TRAIN	1	0.00000	0.00000
2	Neural	_Baseline_	BAD	TRAIN	1	1.00000	1.00000
3	Neural	_Baseline_	BAD	VALIDATE	1	0.00000	0.00000
4	Neural	_Baseline_	BAD	VALIDATE	1	1.00000	1.00000
5	Neural	_Baseline_	BAD	TEST	1	0.00000	0.00000
6	Neural	_Baseline_	BAD	TEST	1	1.00000	1.00000
7	Neural	Neural	BAD	TRAIN	1	0.00000	0.08000
8	Neural	Neural	BAD	TRAIN	1	0.00000	0.09263
9	Neural	Neural	BAD	TRAIN	1	0.00000	0.10737
10	Neural	Neural	BAD	TRAIN	1	0.00000	0.11368
11	Neural	Neural	BAD	TRAIN	1	0.00000	0.12000
12	Neural	Neural	BAD	TRAIN	1	0.00000	0.12421
13	Neural	Neural	BAD	TRAIN	1	0.00000	0.12632
14	Neural	Neural	BAD	TRAIN	1	0.00000	0.13263
15	Neural	Neural	BAD	TRAIN	1	0.00000	0.13474
16	Neural	Neural	BAD	TRAIN	1	0.00000	0.14737
17	Neural	Neural	BAD	TRAIN	1	0.00052	0.15579
18	Neural	Neural	BAD	TRAIN	1	0.00052	0.16632
19	Neural	Neural	BAD	TRAIN	1	0.00052	0.18316
20	Neural	Neural	BAD	TRAIN	1	0.00210	0.19579
21	Neural	Neural	BAD	TRAIN	1	0.00315	0.20000
22	Neural	Neural	BAD	TRAIN	1	0.00315	0.20842
23	Neural	Neural	BAD	TRAIN	1	0.00367	0.21263

Obs	True Positive	True Negative	False Positive	False Negative	First In Group	Last In Group	Up Posterior	Low Posterior
1	357	0	1433	0	0.01638	0.01414	0.01649	0.00000
2	357	0	1433	0	0.01638	0.01414	0.01649	0.00000
3	357	0	1433	0	0.01638	0.01414	0.01649	0.00000
4	357	0	1433	0	0.01638	0.01414	0.01649	0.00000
5	357	0	1433	0	0.01638	0.01414	0.01649	0.00000
6	357	0	1433	0	0.01638	0.01414	0.01649	0.00000
7	38	1907	0	437	0.99957	0.99066	1.00000	0.98994
8	44	1907	0	431	0.98922	0.98331	0.98994	0.98054
9	51	1907	0	424	0.97777	0.96834	0.98054	0.96766
10	54	1907	0	421	0.96698	0.96111	0.96766	0.95897
11	57	1907	0	418	0.95682	0.94955	0.95897	0.94757
12	59	1907	0	416	0.94559	0.93612	0.94757	0.93349
13	60	1907	0	415	0.93087	0.93087	0.93349	0.92554
14	63	1907	0	412	0.92020	0.91320	0.92554	0.91079
15	64	1907	0	411	0.90838	0.90838	0.91079	0.90047
16	70	1907	0	405	0.89257	0.88287	0.90047	0.88271
17	74	1906	1	401	0.88255	0.87657	0.88271	0.87213
18	79	1906	1	396	0.86769	0.86015	0.87213	0.85448
19	87	1906	1	388	0.84880	0.84035	0.85448	0.83941
20	93	1903	4	382	0.83848	0.83108	0.83941	0.82600
21	95	1901	6	380	0.82092	0.81388	0.82600	0.81205
22	99	1901	6	376	0.81022	0.80096	0.81205	0.79862
23	101	1900	7	374	0.79628	0.78792	0.79862	0.78480

The ROC curves in Figure 9.6 are identical to the ones in the second panel of Figure 9.5 and are generated by PROC GPLOT using the following statements. You can exercise more control on the aesthetic and functional aspects of the curve with PROC GPLOT.

```
axis1 length=12cm order=0 to 1 by 0.2 label=(f=swissb h=2)
value=(font=swissb h=2);
axis2 length=12cm order=0 to 1 by 0.2 label=(a=90 f=swissb h=2)
value=(font=swissb h=2);

symbol1 v=none i=join w=2 c=black l=33 h=1.5;
symbol2 v=none i=join w=2 c=black l=1 h=1.5;
symbol3 v=none i=join w=2 c=black l=3 h=1.5;
symbol4 v=none i=join w=2 c=black l=9 h=1.5;


legend1 position=(inside right bottom) frame across=1;


proc gplot data=em(where=(datarole='VALIDATE'));
   plot sensitivity*oneminusspecificity=model / haxis=axis1
vaxis=axis2 legend=legend1;
run;
quit;
```

Figure 9.6 ROC Curves for the Validation Data Using PROC GPLOT

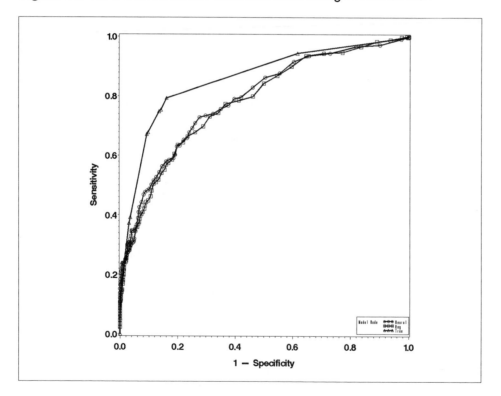

An Introduction to PROC NLMIXED

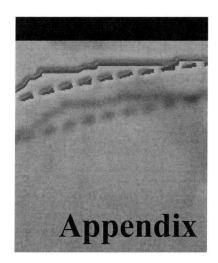

Appendix

PROC NLMIXED plays a central role in this book. Chapters 3, 4, and 5 use it to fit the binormal model, which is one of the fundamental tools to construct and analyze ROC curves.

PROC NLMIXED offers unique and wide-ranging capabilities, but its syntax is different from most other SAS/STAT modeling procedures. For this reason, the PROC NLMIXED code may be difficult to discern and harder to grasp for readers who are used to gaining insight by studying the programs used for model fitting. This appendix is written for users who are familiar with the general SAS/STAT modeling syntax but new to PROC NLMIXED. Regular users of the following procedures may find this appendix helpful: GLM, ANOVA, REG, GENMOD, LOGISTIC, PROBIT, and MIXED.

This appendix is not a summary of all PROC NLMIXED features. Rather, it focuses on capabilities related to ROC curves. SAS/STAT manuals offer exhaustive coverage of PROC NLMIXED and can be helpful for those who want to use this procedure beyond what is covered here.

A.1 Fitting a Simple Linear Model: PROC GLM vs PROC NLMIXED

Consider the following programming statements, which generate a data set of 100 observations. There is a group variable X, taking values of 0 or 1, and an outcome variable Y, which follows a normal distribution within each group. The goal is to fit the following model to this data set using PROC GLM and PROC NLMIXED:

$$y = \alpha + \beta X + \varepsilon$$

where ε has a normal distribution with mean 0 and variance σ^2.

```
data appendix;
  do i=1 to 100;
    x=rand('bernoulli',0.5);
    e=rannor(0);
    y=x+e;
    output;
  end;
run;
```

The following statements fit this model in PROC GLM. The MODEL statement in PROC GLM is specifically designed to recognize these special features of this model:

- Presence of an intercept (there is an option for excluding the intercept)
- Linearity of the model term
- Normal distribution of the error terms with constant variance.

For these reasons, it is sufficient to write `model y=x`.

```
title "Simple Linear Model Using GLM";
proc glm;
  model y=x;
run;
quit;
```

Output A.1 shows the results from PROC GLM. The parameter estimates for α (labeled Intercept) and β (labeled X) are -0.241 and 1.254, with standard errors of 0.148 and 0.197. An estimate of σ^2 is provided by the mean squared error (0.9578).

Output A.1

```
Simple Linear Model Using GLM

The GLM Procedure

Number of Observations Read          100
Number of Observations Used          100

Dependent Variable: y
```

Source	DF	Sum of Squares	Mean Square	F Value	Pr > F
Model	1	38.7439429	38.7439429	40.45	<.0001
Error	98	93.8638957	0.9577949		
Corrected Total	99	132.6078386			

R-Square	Coeff Var	Root MSE	y Mean
0.292169	212.2180	0.978670	0.461163

Source	DF	Type I SS	Mean Square	F Value	Pr > F
x	1	38.74394287	38.74394287	40.45	<.0001

(*continued*)

Output A.1 (*continued*)

Source	DF	Type III SS	Mean Square	F Value	Pr > F
x	1	38.74394287	38.74394287	40.45	<.0001

Parameter	Estimate	Standard Error	t Value	Pr > |t|
Intercept	-0.241051559	0.14754004	-1.63	0.1055
x	1.253953858	0.19715868	6.36	<.0001

PROC NLMIXED does not have a MODEL statement with these features. Rather, the model needs to be communicated to PROC NLMIXED step by step, with specific instructions on the model and the distribution of the variables. The following PROC NLMIXED code fits the same model:

```
title "Simple Linear Model Using NLMIXED";
proc nlmixed;
  parms var=1 alpha=0 beta=0;
  mu=alpha+x*beta;
  model y~normal(mu,var);
run;
quit;
```

It is most useful to deconstruct PROC NLMIXED programs starting from the end. The MODEL statement has no resemblance to the one in PROC GLM. The equal sign is replaced by the tilde (~), which is read as "distributed as." Hence, this model statement is specifying that the response variable *Y* has a normal distribution with mean MU and standard deviation VAR. A key difference between PROC GLM syntax and PROC NLMIXED syntax is that in PROC NLMIXED, only *parameters* are allowed on the right-hand side of a MODEL statement, whereas in PROC GLM, only *variables* are allowed.

The statement `mu=alpha+x*beta` connects the model parameters to the data set variables. It specifies that the mean of the normal distribution is related to the covariate *X* in a linear fashion. Most DATA step functions are available for programming in PROC NLMIXED, so you can construct almost any functional relationship between parameters.

Finally, the PARMS statement tells PROC NLMIXED which of the terms are parameters. The PARMS statement is not required. In its absence, PROC NLMIXED assumes that any term that is not assigned a value, either from the data set or from the programming statements within the procedure, is a parameter. In this case, it is easy to check that omitting the PARMS statement does not alter the output. In more complicated models, the PARMS statement helps you keep track of the model parameters. The numbers following the equal sign after the parameter names are the starting values for the numerical search method employed by PROC NLMIXED. They can be safely ignored for the time being until we briefly discuss the implications of using the maximum likelihood in PROC NLMIXED later in this appendix.

The output from this invocation of PROC NLMIXED appears here. The parameter estimates for α, β, and σ^2 are -0.241, 1.254, and 0.9386, with standard errors 0.148, 0.197, and 0.1327. The estimates for α and β are identical to the ones provided by PROC GLM (up to rounding), but the estimates for σ^2 seem to differ slightly (0.9578 vs. 0.9386). This discrepancy results from the different estimation methods used by the two procedures. It is usually small, especially with moderate to large sample sizes. We will discuss this next.

Output A.2

```
Simple Linear Model Using NLMIXED

The NLMIXED Procedure

                            Specifications

Data Set                                    WORK.APPENDIX
Dependent Variable                          y
Distribution for Dependent Variable         Normal
Optimization Technique                      Dual Quasi-Newton
Integration Method                          None


                    Dimensions

Observations Used                  100
Observations Not Used                0
Total Observations                 100
Parameters                           3


                      Parameters

      var        alpha        beta     NegLogLike

       1            0           0      168.83132


                   Iteration History

 Iter     Calls     NegLogLike        Diff      MaxGrad        Slope

    1         2     148.184071     20.64725     18.21043     -535.042
    2         3     142.074894      6.109177    10.23585      -21.7239
    3         6     140.056178      2.018716     7.234336      -8.129
    4         7     139.602406      0.453772     9.640041      -4.14434
    5         8     138.860921      0.741485     3.650731      -1.86665
    6        10     138.744238      0.116683     1.583923      -0.32313
    7        12     138.727766      0.016472     0.114029      -0.03642
    8        14     138.727636      0.00013      0.008173      -0.00027
```

(continued)

Output A.2 (*continued*)

```
Simple Linear Model Using NLMIXED

The NLMIXED Procedure

                        Iteration History

  Iter    Calls    NegLogLike      Diff     MaxGrad     Slope

    9       16     138.727635    7.861E-7   0.000312   -1.62E-6
   10       18     138.727635    2.272E-9   0.000024   -4.4E-9


          NOTE: GCONV convergence criterion satisfied.


            Fit Statistics

  -2 Log Likelihood                277.5
  AIC (smaller is better)          283.5
  AICC (smaller is better)         283.7
  BIC (smaller is better)          291.3


                                        Parameter Estimates


                    Standard
  Parameter  Estimate   Error    DF   t Value  Pr > |t|   Lower    Upper   Gradient

  var          0.9386   0.1327   100    7.07    <.0001    0.6753  1.2020   -8.19E-6
  alpha       -0.2411   0.1461   100   -1.65    0.1020   -0.5308  0.04872  -0.00002
  beta         1.2540   0.1952   100    6.42    <.0001    0.8667  1.6412   -0.00002
```

PROC GLM uses the method of least squares to find the estimates. Least squares equations for a given statistical linear model have closed form solutions; that is, you can solve a set of linear equations to obtain the estimates. The information in the PROC GLM output such as the sum of squares and mean squares are all related to the analysis of variance decomposition of the total sum of squares, a concept closely connected to the least squares methodology.

In contrast, PROC NLMIXED uses likelihood-based methods. These methods are very general. They can be used effectively for nonlinear or nonnormal models. The price of the generality is the lack of closed-form solutions. In general, likelihood equations are nonlinear and can only be solved by numerical search algorithms. For example, under the Specifications heading in the PROC NLMIXED output, the method of optimization is listed as Dual Quasi-Newton. This is one method of solving the likelihood equations to obtain the estimates. The Dual Quasi-Newton algorithm (and others used in PROC NLMIXED, which can be specified by the METHOD option in the procedure statement) is *iterative*. The algorithm takes small steps, called *iterations*, toward the solution. The starting point of the iterations are the ones listed (after the equal signs) in the PARAMS statement. The listing under Iteration History in the output gives details of each such iteration. This table is normally useful only to advanced users of PROC NLMIXED.

The reason for the slight difference in the estimates of σ^2 is that the maximum likelihood estimates use n (total sample size) as the denominator for the variance estimates, whereas the least squares methods use $n-p$, where p is the number of model parameters. To the extent that p is small compared to n, the estimates will be close. You can obtain the maximum likelihood estimate from the PROC GLM output by dividing the error sum of squares (93.8639) by 100.

Model fit is assessed by likelihood-based criteria such as Akaike's Information Criterion (AIC) or Bayes Information Criterion (BIC) in PROC NLMIXED, which are reported under Fit Statistics. These are useful for model comparison. Users of PROC MIXED, another procedure that used likelihood methods, will be familiar with these criteria. In contrast, model fit in PROC GLM is usually assessed by the omnibus *F*-test, which is a function of the sum of squares.

The structure of the Parameter Estimates table is similar to the output of other SAS/STAT programs. There is normally another column labeled Alpha in the output, reporting the significance level used in the confidence intervals, which is removed here to save space. An additional piece of information, the value of the gradient at the final iteration, is also provided but is likely to benefit only advanced users.

A.2 PROC NLMIXED and the Binormal Model

It may be useful to understand how the PROC NLMIXED programs used in Chapters 3 and 5 are developed. First, consider the model and code from Section 3.5:

```
proc nlmixed;
  parms m1=0 m0=0 s1=1 s0=1;
  if gold=1 then m=m1;else if gold=0 then m=m0;
  if gold=0 then s=s1**2;else if gold=0 then s=s0**2;
  a=(m1-m0)/s1;
  b=s0/s1;
  model y~normal(m,s);
  estimate 'a' a;
  estimate 'b' b;
  estimate 'AUC' probnorm(a/sqrt(1+b**2)));
run;
quit;
```

The MODEL statement specifies a normally distributed outcome with mean M and variance S, which are defined by the two IF statements. This would produce an ROC analysis, but you would have to manually (or in a DATA step) generate point estimates for the binormal parameters a and b, as well as the area under the ROC curve (AUC). Obtaining their variances would take considerable time and some knowledge of PROC IML. This underscores the practicality of the ESTIMATE statement, where you can define any nonlinear transformation using a DATA step-like programming approach and obtain the point estimates and standard errors. The following code shows two different ways of doing so:

1. These quantities can be defined first and then estimated. This method was used for the binormal parameters a and b in the code above.

2. The definition can be performed in the ESTIMATE step. This method was used for the AUC in the code above.

Note that the ESTIMATE statement in PROC GLM and other SAS/STAT procedures works only with linear transformation, so obtaining estimates of a, b, and AUC is not possible with, say, PROC GLM.

Now let's revisit the latent binormal model from Chapter 5. This highlights a major strength of PROC NLMIXED: the ability to specify an arbitrary likelihood function.

```
proc nlmixed data=moody gconv=0;
  parms alpha=1 theta1=1 theta2=2 theta3=3 beta=1;
  bounds theta1>0, theta2>0, theta3>0;
  eta1=alpha*def;
  eta2=exp(beta*def);
        if r11=1 then z=probnorm(-eta1/eta2);
  else if r11=2 then z=probnorm((theta1-eta1)/eta2)-
        probnorm(-eta1/eta2);
  else if r11=3 then z=probnorm((theta2+theta1-eta1)/eta2)-
        probnorm((theta1-eta1)/eta2);
  else if r11=4 then z=probnorm((theta3+theta2+theta1-eta1)/eta2)-
        probnorm((theta2+theta1-eta1)/eta2);
  else if r11=5 then z=1 - probnorm((theta3+theta2+theta1-
        eta1)/eta2);
  if z>1e-6 then ll=log(z);
    else ll=-1e6;
  model r11 ~ general(ll);
  estimate 'AUC' probnorm(alpha/sqrt(1+beta**2));
run;
```

The MODEL statement specifies a GENERAL (LL) model. This means that the (logarithm of the) likelihood function is being defined by the user through term LL. Recall that R11 represents the ordinal credit scores and the underlying model is a latent ordinal regression model. Each IF statement specifies the contribution to the likelihood from every possible value of R11, the ordinal ratings. These contributions were derived in Chapter 5 to be

$$\gamma_k(z) = \Phi\left(\frac{\theta_k - \alpha D}{e^{\beta z}}\right)$$

The structure of the IF statements mimics this formulation closely. These statements uniquely define a value for *Z*, the likelihood, and hence LL, which is the logarithm of *Z*. The condition that the contributions to the likelihood that are less than 10^{-6} will be counted as 10^{-6} ensures that very small contributions do not cause numerical instability.

References

Allison, P.D. (1995). *Survival Analysis Using the SAS System: A Practical Guide*. Cary, NC: SAS Institute Inc.

Cantor, A.B. (2003). *SAS Survival Analysis Techniques for Medical Research, Second Edition*. Cary, NC: SAS Institute Inc.

Cerrito, P. B. (2006). *Introduction to Data Mining Using SAS Enterprise Miner*. Cary, NC: SAS Institute Inc.

Cox, D.R. (1972). "Regression models and life tables (with discussion)." *Journal of the Royal Statistical Society*. Series B 34:187–220.

Cox, D.R. (1975). "Partial likelihood." *Biometrika*. 62:269–276.

DeLong, E.R., D.M. DeLong, and D.L. Clarke-Pearson. (September, 1988). "Comparing the areas under two or more correlated receiver operating characteristic curves: A nonparametric approach." *Biometrics*. 44(3):837–845.

Efron, B., and R. Tibshirani. (1993). *An Introduction to the Bootstrap*. New York: Chapman & Hall.

Gönen, M., and G. Heller. (2005). "Concordance probability and discriminatory power in proportional hazards regression." *Biometrika*. 92:965–970.

Gönen, M., and G. Heller. (2007). "Lehmann family of ROC curves." (Under peer review). Available at www.bepress.com/mskccbiostat/paper11/.

Güttler, A. (2005). "Using a bootstrap approach to rate the raters." *Financial Markets and Portfolio Management*. 19(3):277–295.

Hanley, J.A. (August, 1988). "The robustness of the 'binormal' assumptions used in fitting ROC curves." *Medical Decision Making*. 8(3):197–203.

Harrell, F.E., Jr., R.M. Califf, D.B. Pryor, K.L. Lee, and R.A. Rosati. (1982). "Evaluating the yield of medical tests." *Journal of the American Medical Association*. 247(18):2543–2546.

Heagerty, P.J., T. Lumley, and M.S. Pepe. (2000). "Time-dependent ROC curves for censored survival data and a diagnostic marker." *Biometrics*. 56(2):337–344.

Jarnagin, W.R., M. Gönen, Y. Fong, R.P. DeMatteo, L. Ben-Porat, S. Little, C. Corvera, S. Weber, and L.H. Blumgart. (October, 2002). "Improvement in perioperative outcome after hepatic resection: Analysis of 1,803 consecutive cases over the past decade." *Annals of Surgery*. 236(4):397–407. Discussion 406-7. PMID: 12368667 [PubMed - indexed for MEDLINE].

Lehmann, E.L. (1953). "The power of rank tests." *The Annals of Mathematical Statistics*. 24:23–43.

Littell, R.C., G.A. Milliken, W.W. Stroup, R.D. Wolfinger, and O. Schabenberger. (2006). *SAS for Mixed Models, Second Edition*. Cary, NC: SAS Institute Inc.

Pepe, M.S. (2003). *The Statistical Evaluation of Medical Tests for Classification and Prediction*. New York: Oxford University Press.

Perkins, N.J., and E.F. Schisterman. (April, 2006). "The inconsistency of 'optimal' cutpoints using two criteria based on the Receiver Operating Characteristic Curve." *American Journal of Epidemiology*. 163(7):670–675.

Sakia, R.M. (1992). "The Box-Cox transformation technique: A review." *The Statistician*. 41:169–178.

Stokes, M.E., C.S. Davis, and G.G. Koch. (2000). *Categorical Data Analysis Using the SAS System, Second Edition.* Cary, NC: SAS Institute Inc.

Thornes, J.E., and D.B. Stephenson. (September, 2001). "How to judge the quality and value of weather forecast products." *Meteorological Applications.* 8(3):307–314.

Tosteson, A.N., and C.B. Begg. (1988). "A general regression methodology for ROC curve estimation." *Medical Decision Making.* 8(3):204–215.

Wong, R.J., D.T. Lin, H. Schöder, S.G. Patel, M. Gönen, S. Wolden, D.G. Pfister, J.P. Shah, S.M. Larson, and D.H. Kraus. (October, 2002). "Diagnostic and prognostic value of [^{18}F]fluorodeoxyglucose positron emission tomography for recurrent head and neck squamous cell carcinoma." *Journal of Clinical Oncology.* 20(20):4199–4208.

Youden, W.J. (1950). "Index for rating diagnostic tests." *Cancer.* 3(1):32–35.

Zajick, D.C., Jr., W.B. Morrison, M.E. Schweitzer, J.A. Parellada, and J.A. Carrino. (November, 2005). "Benign and malignant processes: Normal values and differentiation with chemical shift MR imaging in vertebral marrow." *Radiology.* 237(2):590–596.

Zhou, X-H., D.K. McClish, and N.A. Obuschowski. (2002). *Statistical Methods in Diagnostic Medicine.* New York: John Wiley & Sons.

Index

Books Available from SAS Press

Advanced Log-Linear Models Using SAS®
by **Daniel Zelterman**

Analysis of Clinical Trials Using SAS®: A Practical Guide
by **Alex Dmitrienko, Geert Molenberghs, Walter Offen,** and
Christy Chuang-Stein

Analyzing Receiver Operating Characteristic Curves with SAS®
by **Mithat Gönen**

Annotate: Simply the Basics
by **Art Carpenter**

*Applied Multivariate Statistics with SAS® Software,
Second Edition*
by **Ravindra Khattree**
and **Dayanand N. Naik**

*Applied Statistics and the SAS® Programming Language,
Fifth Edition*
by **Ronald P. Cody**
and **Jeffrey K. Smith**

An Array of Challenges — Test Your SAS® Skills
by **Robert Virgile**

*Building Web Applications with SAS/IntrNet®: A Guide to the
Application Dispatcher*
by **Don Henderson**

*Carpenter's Complete Guide to the SAS® Macro Language,
Second Edition*
by **Art Carpenter**

Carpenter's Complete Guide to the SAS® REPORT Procedure
by **Art Carpenter**

The Cartoon Guide to Statistics
by **Larry Gonick**
and **Woollcott Smith**

*Categorical Data Analysis Using the SAS® System,
Second Edition*
by **Maura E. Stokes, Charles S. Davis,**
and **Gary G. Koch**

Cody's Data Cleaning Techniques Using SAS® Software
by **Ron Cody**

*Common Statistical Methods for Clinical Research with
SAS® Examples, Second Edition*
by **Glenn A. Walker**

The Complete Guide to SAS® Indexes
by **Michael A. Raithel**

*CRM Segmentation and Clustering Using SAS® Enterprise
Miner™*
by **Randall S. Collica**

*Data Management and Reporting Made Easy with
SAS® Learning Edition 2.0*
by **Sunil K. Gupta**

Data Preparation for Analytics Using SAS®
by **Gerhard Svolba**

*Debugging SAS® Programs: A Handbook of Tools and
Techniques*
by **Michele M. Burlew**

*Decision Trees for Business Intelligence and Data Mining: Using
SAS® Enterprise Miner™*
by **Barry de Ville**

*Efficiency: Improving the Performance of Your SAS®
Applications*
by **Robert Virgile**

The Essential Guide to SAS® Dates and Times
by **Derek P. Morgan**

The Essential PROC SQL Handbook for SAS® Users
by **Katherine Prairie**

*Fixed Effects Regression Methods for Longitudinal Data
Using SAS®*
by **Paul D. Allison**

Genetic Analysis of Complex Traits Using SAS®
Edited by **Arnold M. Saxton**

A Handbook of Statistical Analyses Using SAS®, Second Edition
by **B.S. Everitt**
and **G. Der**

Health Care Data and SAS®
by **Marge Scerbo, Craig Dickstein,**
and **Alan Wilson**

The How-To Book for SAS/GRAPH® Software
by **Thomas Miron**

*In the Know ... SAS® Tips and Techniques From
Around the Globe, Second Edition*
by **Phil Mason**

Instant ODS: Style Templates for the Output Delivery System
by **Bernadette Johnson**

*Integrating Results through Meta-Analytic Review Using
SAS® Software*
by **Morgan C. Wang**
and **Brad J. Bushman**

Introduction to Data Mining Using SAS® Enterprise Miner™
by **Patricia B. Cerrito**

Learning SAS® by Example: A Programmer's Guide
by **Ron Cody**

support.sas.com/pubs

SAS® Macro Programming Made Easy, Second Edition
by **Michele M. Burlew**

SAS® Programming by Example
by **Ron Cody**
and **Ray Pass**

SAS® Programming for Researchers and Social Scientists, Second Edition
by **Paul E. Spector**

SAS® Programming in the Pharmaceutical Industry
by **Jack Shostak**

SAS® Survival Analysis Techniques for Medical Research, Second Edition
by **Alan B. Cantor**

SAS® System for Elementary Statistical Analysis, Second Edition
by **Sandra D. Schlotzhauer**
and **Ramon C. Littell**

SAS® System for Regression, Third Edition
by **Rudolf J. Freund**
and **Ramon C. Littell**

SAS® System for Statistical Graphics, First Edition
by **Michael Friendly**

The SAS® Workbook and *Solutions* Set
(books in this set also sold separately)
by **Ron Cody**

Saving Time and Money Using SAS®
by **Philip R. Holland**

Selecting Statistical Techniques for Social Science Data: A Guide for SAS® Users
by **Frank M. Andrews, Laura Klem, Patrick M. O'Malley, Willard L. Rodgers, Kathleen B. Welch,**
and **Terrence N. Davidson**

Statistical Quality Control Using the SAS® System
by **Dennis W. King**

Statistics Using SAS® Enterprise Guide®
by **James B. Davis**

A Step-by-Step Approach to Using the SAS® System for Factor Analysis and Structural Equation Modeling
by **Larry Hatcher**

A Step-by-Step Approach to Using SAS® for Univariate and Multivariate Statistics, Second Edition
by **Norm O'Rourke, Larry Hatcher,**
and **Edward J. Stepanski**

Step-by-Step Basic Statistics Using SAS®: Student Guide
and *Exercises*
(books in this set also sold separately)
by **Larry Hatcher**

Survival Analysis Using SAS®: A Practical Guide
by **Paul D. Allison**

Tuning SAS® Applications in the OS/390 and z/OS Environments, Second Edition
by **Michael A. Raithel**

Univariate and Multivariate General Linear Models: Theory and Applications Using SAS® Software
by **Neil H. Timm**
and **Tammy A. Mieczkowski**

Using SAS® in Financial Research
by **Ekkehart Boehmer, John Paul Broussard,**
and **Juha-Pekka Kallunki**

Using the SAS® Windowing Environment: A Quick Tutorial
by **Larry Hatcher**

Visualizing Categorical Data
by **Michael Friendly**

Web Development with SAS® by Example, Second Edition
by **Frederick E. Pratter**

Your Guide to Survey Research Using the SAS® System
by **Archer Gravely**

JMP® Books

Elementary Statistics Using JMP®
by **Sandra D. Schlotzhauer**

JMP® for Basic Univariate and Multivariate Statistics: A Step-by-Step Guide
by **Ann Lehman, Norm O'Rourke, Larry Hatcher,**
and **Edward J. Stepanski**

JMP® Start Statistics, Third Edition
by **John Sall, Ann Lehman,**
and **Lee Creighton**

Regression Using JMP®
by **Rudolf J. Freund, Ramon C. Littell,**
and **Lee Creighton**